W9-CUK-335

Dynamic Characteristics of Ion-Selective Electrodes

Authors

Ernö Lindner, Ph.D.
Research Scientist
Institute for General and
Analytical Chemistry
Technical University
Budapest, Hungary

Klára Tóth, Ph.D.
Associate Professor
Institute for General and
Analytical Chemistry
Technical University
Budapest, Hungary

Ernö Pungor, Ph.D.
Head of Institute
Institute for General and
Analytical Chemistry
Technical University
Budapest, Hungary

CRC Press, Inc.
Boca Raton, Florida

Library of Congress Cataloging-in-Publication Data

Lindner, Ernö.
Dynamic characteristics of ion-selective electrodes.

Bibliography: p.
Includes index.
1. Electrodes, Ion selective. I. Tóth, Klára.
II. Pungor, E. (Ernö) III. Title
QD571.P86 1988 541.3′724 87-20949
ISBN 0-8493-6493-0

This book represents information obtained from authentic and highly regarded sources. Reprinted material is quoted with permission, and sources are indicated. A wide variety of references are listed. Every reasonable effort has been made to give reliable data and information, but the author and the publisher cannot assume responsibility for the validity of all materials or for the consequences of their use.

Direct all inquiries to CRC Press, Inc., 2000 Corporate Blvd., N.W., Boca Raton, Florida, 33431.

International Standard Book Number 0-8493-6493-0

Library of Congress Card Number 87-20949
Printed in the United States

PREFACE

Since the introduction of ion-selective electrodes, a continuing objective of research has been characterization of their dynamic properties. In modern applications, the speed influences precision and reproducibility of analysis, especially in flow-through systems used in process control and clinical applications. New sensor response dynamics must be measured and compared with other available types. The transient responses, even for dip-type applications, are important for the detection of fouling and an indication of limited future lifetime.

Furthermore, various demands for analytical procedures employing ISEs (e.g., enhancement of accuracy, selectivity, and rate of analysis) made necessary the detailed study of the individual steps in the electrode process, i.e., the elucidation of the working mechanism of ion-selective electrodes. These demands gave further impetus to the application of transient techniques in ISE studies. In any relaxation or transient technique, the system in equilibrium or steady state is perturbed by a pulse or periodical signal, and the relaxation processes leading to the new equilibrium are analyzed to obtain quantitative data on the rates of ''fast'' processes. The experimental techniques most frequently employed in ion-selective membrane electrode research are the impedance method, the polarization studies applying galvanostatic current step, and the so-called activity step method.

The present book deals with the principle of the aforementioned techniques and discusses the information they provide for electrode kinetics. Special attention is paid to the activity step method, since this technique is carried out under zero current potentiometric conditions and allows the study of the processes at the perturbed membrane-solution interface. Different models developed for the interpretation of the transient potential following an activity step in the absence or in the presence of interfering ions are discussed.

Finally, an attempt is made towards the definition of response time of an ion-selective electrode and an electrochemical cell. The importance of dynamic properties of ion-selective electrodes in practical analysis is demonstrated on selected examples.

The authors wish to acknowledge the comments and suggestions of Dr. T. Garai for reading the manuscript and the great care and attention in translating selected chapters of it. The careful work of Mrs. K. Barabás in typing the manuscript is also gratefully acknowledged.

THE AUTHORS

Ernö Lindner, M.Sc., Ph.D., C.Ch.Sci., is a senior scientist in the Institute for General and Analytical Chemistry, Technical University of Budapest, Hungary.

Dr. Ernö Lindner graduated in 1971 from the Faculty of Chemical Engineering of the Technical University, Budapest. He obtained his Ph.D. degree from the Technical University of Budapest and received a Candidate of Chemical Sciences (C.Ch.Sci.) degree from the Hungarian Academy of Sciences in 1983.

Dr. Lindner is a member of the Electroanalytical Commission of the Hungarian Academy of Sciences, the secretary of the electroanalytical subdivision of the Hungarian Chemical Society, and a national representative of the IUPAC Commission V. 5 on Electroanalytical Chemistry.

Dr. Lindner worked 1 year (1974 to 1975) in Zürich at the ETH in Prof. W. Simon's laboratories on ionophore-based electrodes, spent 3 months in Prof. R. Bates and R. P. Buck's laboratories in 1981 and 1986, respectively, dealing with problems of the determination of activity coefficients and membrane impedances. His current research interests include the design of new ionophores and studies on the working mechanism of different types of ion-selective electrodes as well as the application of ISEs and other electroanalytical sensors.

He has published over 35 scientific papers and is co-author of several reviews on ion-selective electrodes.

Klára Tóth, M.Sc., Ph.D., C.Ch.Sc., is an associate professor in the Institute for General and Analytical Chemistry, at the Technical University of Budapest, Hungary.

Dr. Tóth graduated in 1962 from the Faculty of Pharmacy, Semmelweis Medical School, Budapest, with a degree in pharmacy and obtained her Ph.D. degree in 1964 from the University of Chemical Industries, Veszprém, Hungary. In 1970, she received a C.Sc. degree from the Hungarian Academy of Sciences.

Dr. Tóth is a member of the Analytical Division of the Hungarian Academy of Sciences and the Promotion Committee of the Hungarian Academy of Sciences. She is the secretary of the Electroanalytical Commission of the Hungarian Academy of Sciences, the president of the Electroanalytical Commission of the Hungarian Chemical Society, and an Associate-Member of the IUPAC Commission on Electroanalytical Chemistry. In 1978 she received the award of the Hungarian Academy of Sciences.

Her current research interests include the design of new ion-sensitive materials, the study of bulk and interfacial processes of liquid membranes, and the development of automated electrochemical techniques for the analysis of biological materials.

She has had over 100 scientific papers published and is co-author of a work entitled *Indicator Electrodes as Sensors for Process Control,* and also has had several reviews and chapters published on ion-selective electrodes.

Professor Dr. Ernö Pungor was born October 30, 1923, in Vasszécsény, Hungary. He ended his secondary school in Szombathley, finishing his studies in chemistry in Budapest in 1948. A year later he got his Ph.D., and in 1952 became candidate of chemical sciences and doctor of chemical sciences in 1956. In 1967, he was elected corresponding member of the Hungarian Academy of Sciences and in 1976 he became member of the Hungarian Academy of Sciences.

After finishing his university studies he worked in the Institute for Inorganic and Analytical Chemistry at the Eötvös University in Budapest, first as a lecturer, and later as senior lecturer and reader. He was appointed professor in 1962 to the Department of Analytical Chemistry at the University Veszprém. Since 1970 he has been head of the Institute for General and Analytical Chemistry of the Technical University in Budapest.

Professor Pungor's main scientific activities are in the field of instrumental analysis, notably in electrochemistry. During the 1950s, he did scientific work in oscillometry, flame photometry, and atomic absorption, and in the 1960s, he developed precipitate based ion-selective electrodes. In the 1970s, Pungor developed some new lines of automatic analysis. He wrote four monographs and chapters in many other monographs. He has written 400 scientific publications.

He is editor of the Hungarian Scientific Instruments, the Hungarian Chemical Journal, and a member of many editorial boards in the field of analytical chemistry in Hungary and abroad. Dr. Pungor is president of the European Association of the Editors of Chemical Journals and chairman of the working party of analytical chemistry of the Federation of European Chemical Societies.

TABLE OF CONTENTS

Chapter 1
History and Development 1

Chapter 2
Transient Techniques Applied to Study Ion-Selective Electrodes 3

I. Introduction 3
II. The Impedance Method 4
  A. The Impedance Plane Plots: Definitions 4
  B. Application of the Method for Ion-Selective Electrode Studies 8
  C. Information Obtained from Impedance Plane Plots 8
III. Polarization Studies of Ion-Selective Electrodes 11
  A. Methods for Determining Exchange Current Densities 11
  B. A Mixed Potential Ion-Selective Electrode Theory 16
IV. Studies Indirectly Related to the Dynamic Characteristics of Ion-Selective Electrodes 19
V. The Activity Step Method 21
  A. The Principle of the Method 21
  B. Experimental Techniques 22
    1. The Dipping (Immersion) Method 22
    2. The Injection Method 22
    3. Special Techniques 23
  C. Parameters Affecting the Transient Function of Ion-Selective Electrodes 26
    1. The Primary Ion Concentration 27
    2. Interferences 28
    3. Hydrodynamic Conditions 28
    4. Membrane Bulk and Surface Parameters 30
VI. Comparison of the Different Techniques 31

Chapter 3
Theories of Transient Potentials (Following an Activity Step in the Absence of Interfering Ions) 33

I. Introduction 33
II. Diffusion Through a Stagnant Layer 35
  A. Limits of the Diffusion Model 44
    1. The Effect of the Activity Level 49
    2. The Effect of the Activity Ratios 51
    3. Conclusions 54
  B. The Multielectrode Model 54
III. Kinetics of Interfacial Reactions 60
  A. Energy Barrier Concept 60
  B. First-Order Chemical Kinetics and the Consecutive Reaction Model 62
  C. Second-Order Chemical Kinetics 64
IV. Diffusion Within the Ion-Sensing Membrane 68

V. Unified Models for Transient Functions 77

Chapter 4
Transient Potentials in the Presence of Interfering Ions 81

I. Introduction 81
II. The Segmented Membrane Model 84
III. Differences in the Surface Activity Due to Adsorption/Desorption Processes 90
A. Qualitative Interpretation 90
1. Effect of Flow Rate 94
2. Effect of Interfering Ion Activity 94
3. Effect of the Direction of Activity Change 95
4. Effect of Electrode Surface Conditions 95
5. Effect of Temperature on the Transient Signal 96
B. Quantitative Interpretation 97
IV. Comparison of the Different Models 104

Chapter 5
Determination and Definition of Response Time 107

I. Introduction 107
II. Dynamic Response of an Ion-Selective Electrode and an Electrochemical Cell 107
III. Definition of Response Time 108
IV. Estimation of the Equilibrium Potential 112
V. Selected Examples 114
A. Interpretation of the Data of Mertens et al. 116
B. Interpretation of the Data of Denks and Neeb 116

Chapter 6
The Importance of Dynamic Properties of Ion-Selective Electrodes in Practical Analysis 121

References 125

Index 131

Chapter 1

HISTORY AND DEVELOPMENT

The working mechanism of ion-selective electrodes has been the subject of scientific research since the first decade of the 20th century, when the first glass electrode was developed. At first, the potential response was interpreted in terms of Donnan equilibrium. Later, in the 1930s, Nicolsky initiated a remarkable development with his new theory which assumed that the electrode response is dependent on the active sites of the glass, capable of ion exchange. Equilibrium potentials were derived from electrochemical potentials and the Nicolsky concept also enabled the selectivity coefficient of an electrode to be calculated.

Further development in the theories of electrode potentials has led to the interpretation of the response functions of semipermeable membranes. According to some of the new theories, the concept of selective ion transport has been applied to understand the electrode response mechanism. In other theories, the electrode potential has been ascribed to space charge (charge separation) on the electrode solution interface.

In order to understand the mechanism of the ion-selective electrodes, it would be ideal to study the various reactions producing the electrochemical signal. However, it is very difficult to define all of the reactions that determine the overall electrode reaction.

The first step of the potential determining reaction is the transport of the ions from the bulk of the solution to the surface of the electrode. In nonstirred solution, only the chemical potential difference is responsible for the transport, while in stirred solution, the convective diffusion model has to be applied.

At the electrode surface, the ion may enter in various reaction steps. If the ion approaches the electrode surface at an inactive site, surface diffusion or back diffusion to the bulk of the solution may take place. The next process of the overall electrode reaction is charge transfer as the primary ion crosses the interface to participate in an ion exchange process. However, prior to a surface ion exchange, the ion must be desolvated and gotten into an energetically favorable state. All of the theories agree as long as these processes are concerned.

In terms of the theory involving ion transfer, the ion, after entering the electrode phase, moves towards the place of lower activities, due to chemical potential decrease. On the other hand, in terms of the space charge theory, a charge separation takes place at the membrane solution interface and is characterized by different distribution functions in the two contacting faces. If the potential determining ion is present on both sides of the ion-selective electrode membrane, similar phenomena take place on both surfaces.

Intensive research work on ion-selective electrodes started in the early 1960s. The study of the effect of phase composition on the equilibrium potentials and of reactions leading to potential establishment started in our institute, also in the 1960s. It was clear that equilibrium potentials as predicted by thermodynamics cannot provide information concerning the nature of partial processes leading to the equilibrium state.

It was kept in mind that electrode studies under dynamic conditions may provide information on the partial processes of the electrode response mechanism. Consequently, in the second half of the 1960s, great hopes were attached to response time measurements using the activity step method. For this purpose, we introduced the switched wall-jet technique. Metal electrodes, for which kinetic information was already available, were studied first, and later, other electrode types, mainly halide-selective electrodes, were investigated. However, the experiments have revealed that the information obtained is less valuable than first expected. Diffusion in the solution, as a rate-controlling process, has covered up other, faster reactions in most cases. Hence, in the 1970s, our interest was directed to study the initial part of dynamic response curves, which were supposed to give information on other kinetic factors, too. In view of the experiments, the experimental conditions could be selected under which overlapping effect of diffusion can be neglected.

In the late 1970s and early 1980s, we began to focus our attention on a new aspect of the dynamic behavior of the electrodes, namely, the response in the presence of interfering ions. The ratio of the activity of the primary ion to that of the interfering ion was around that allowed by the selectivity coefficient value according to the Nicolsky equation. These studies provided another insight into some fundamental phenomena affecting the electrode response mechanism.

The activity step method, however, is not the only possibility of dynamic response studies. Important information may be obtained on the electrode mechanism by studying relaxation processes after an applied voltage or current signal. A further source of information is offered by the use of an AC impedance method. In the latter case, resistance terms and time constants can be determined from complex plane diagrams.

The material in this book is based on the literature of ion-selective electrodes and on our unpublished results. It was our objective to give guidance to the readers concerning the dynamic investigations of ion-selective electrodes and the methods of evaluation and interpretation of experimental results.

The dynamic properties of ion-selective electrodes are of great practical importance. In all analytical applications, especially in flow analytical techniques, the knowledge of the dynamic behavior of ion-selective electrodes is of utmost importance. In the majority of cases, the time requirement of convective diffusion under flow conditions is commensurable with the response time of the sensors. However, the response time of some electrodes is much longer than the time required by convective diffusion. With the use of these electrodes, the distortion due to the long response time has to be considered in the design of the experimental conditions and in the evaluation of the results. Some monitors provide erroneous data because of the lack of knowledge concerning the response time. This is especially true with closed-loop reactors based on enzyme electrodes, which are characterized by rather long response times. Ion-selective electrodes are also applicable to theoretical physicochemical studies, e.g., to follow oscillating reactions.

In reaction kinetic applications, it is very important to consider the possible distortion due to slow electrode response. If this fact is not taken into account, the kinetic parameters will be determined with an error. Some practical applications are also mentioned in the book.

Chapter 2

TRANSIENT TECHNIQUES APPLIED TO STUDY ION-SELECTIVE ELECTRODES

I. INTRODUCTION

To study electrochemical reactions, the relaxation or transient techniques have proved to be particularly advantageous. When applying these techniques, the system in equilibrium or steady state is perturbed by a pulse or periodical signal and the relaxation processes (i.e., the response as a function of time) resulting in a new equilibrium or steady state are analyzed to extract the desired kinetic information.[1]

To disturb the studied system or reaction as a perturbing signal, a current, or voltage step, or periodical signals are applied most frequently.[1-3] In the study of the dynamic response of ion-selective electrodes, the cell or membrane potential vs. time functions is often recorded after a ''step'' activity change on one side of these membranes.[4-8]

The selection of the experimental techniques depends on the following:[1]

1. The rate of the reaction studied
2. The information required, including accuracy
3. The limitations set by the experimental conditions, such as the solvent properties (conductivity, viscosity, etc.), temperature, and pressure

In studying ion-selective membrane electrodes, the experimental techniques may be grouped according to the nature of information required:[8]

1. Methods providing information on the electrical properties of ion-selective membranes (experiments at $i \neq 0$ current conditions): (α) impedance methods[9-12]; (β) polarization studies[13,14]; (γ) membrane transport experiments[15-17]
2. Methods used for the investigation of individual processes regarded decisive in respect to the dynamic properties of ion-selective electrodes (e.g., the study of extraction kinetics with liquid membranes and the study of adsorption kinetics and determination of the rate of dissolution and of isotope distribution, with precipitate-based ion-selective membranes, etc.)
3. Methods used to study the phenomena at the membrane-solution interface; interpretation and mathematical description of the transient functions following a sudden step change in the ion activity at one side of the membrane; that is, determination of the response time of the electrode is considered to be important from a practical point of view (experiments with $i = 0$)

The different techniques are equally relevant for the interpretation of the response mechanism of ion-selective electrodes. However, a realistic picture can be achieved only by the simultaneous evaluation of all the results obtained by different methods.

Accordingly, efforts have been made to compare the results yielded by different methods.[14,17-25] Thus, with the help of digital simulation, Buck et al.[23] compared the impedance technique and the activity step method, while Buffle and co-workers[24,25] found a close correlation between the dissolution rate of precipitate-based membranes and the rate of potential response following an activity step.

II. THE IMPEDANCE METHOD

The impedance method permits the transport characterization of a system in a braod sense, i.e., the determination of conductivity, capacitance, and inductance, i.e., the elements of the equivalent passive network.

Real systems are represented by equivalent electrical circuits (R,L,C passive networks), whose current-potential-time properties are exactly the same as those of the real system.

The impedance method for the analysis of electrochemical reactions was introduced by Sluyters-Rehbach and Sluyters[26] and found considerable application in the characterization of different electrochemical systems (e.g., see References 27 to 29). The extension of the technique to ion-selective membranes was initiated by Buck.[9-12,22]

A. The Impedance Plane Plots: Definitions

When a sinusoidal AC current or voltage is imposed across an electrochemical cell, the output AC voltage or current, respectively, is characteristic of the system (Figure 1).

In Figure 1, the input signal is an AC current of ω angular frequency and ΔI amplitude:

$$I(\omega t) = \Delta I \sin(\omega t) \tag{1}$$

while the output signal is the alternating voltage of $\Delta\Phi$ amplitude:

$$\Phi(\omega t) = \Delta\Phi \sin(\omega t + \Theta) \tag{2}$$

where $\omega = 2\pi f$ and f is the sinusoidal frequency in c/sec, while Θ is the phase angle shift. Thus, one can define an impedance, Z, with a magnitude as

$$|Z| = \frac{\Delta\Phi}{\Delta I} \tag{3}$$

and a phase angle, θ, which is equal to the phase difference between the applied sinusoidal current and the resultant sinusoidal voltage.

Accordingly, the impedance is a vector quantity, since it has both a magnitude and a direction. Two such component vectors are convenient to represent as points in a complex plane (Figure 2).

A complex impedance vector then can be described by the absolute value of the impedance ($|Z|$) and the phase angle (Θ):

$$|Z| = \sqrt{Z_R^2 + Z_I^2} \tag{4}$$

$$\Theta = \arctan\left(\frac{Z_I}{Z_R}\right) \tag{5}$$

or by its real and imaginary components, Z_R and Z_I projected on the x and y axes. $|Z|$ and Θ are frequency dependent, thus,

$$Z(\omega) = Z_R(\omega) - iZ_I(\omega) \tag{6}$$

where $i = \sqrt{-1}$.

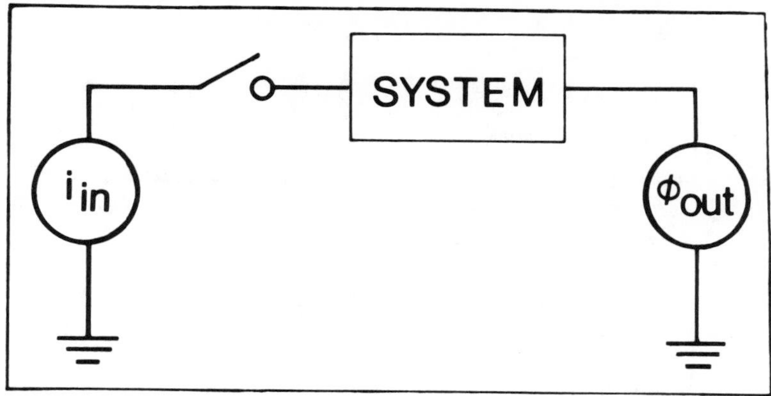

FIGURE 1. "Thought" experiment circuit for step, pulse, or periodic current perturbation of a system. (From Buck, R. P., *Hung. Sci. Instrum.*, 49, 7, 1980. With permission.)

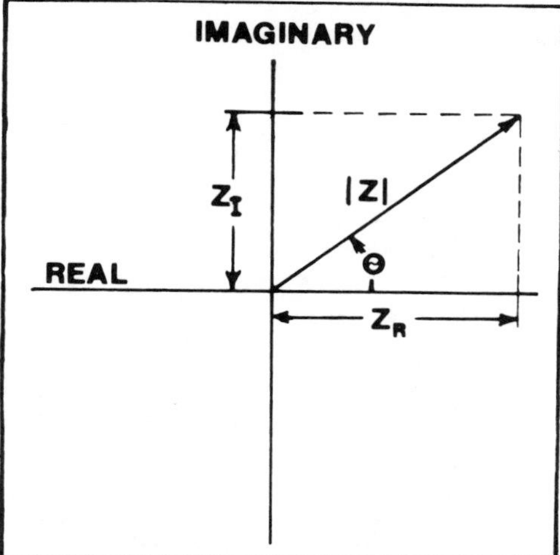

FIGURE 2. Impedance plotted in the complex plane.

The information provided by the impedance method for a real system can be visualized by considering the response of simple circuit elements (Figure 3) and circuits under the experimental condition shown in Figure 1.[12,29]

It is apparent that the ohmic resistance R ($Z_R = R$, $\theta = 0$) is represented by a point on the abscissa at any frequency, while the impedance of capacitance C ($Z_I = 1/\omega C$) is represented by various points of the ordinate ($\theta = 90°$) depending on the frequency.

The equivalent circuits related to electrochemical cells generally consist of serial and parallel RC circuits, thus it is worthwhile to study their impedance spectra (Figure 3c, d).

In the case of a resistance and capacitance in the series:

$$Z = R_s - i/\omega C_s \tag{7}$$

since $Z_R = R_s$ and $Z_I = 1/\omega C_s$, then

$$Z^2 = R_s^2 + \left(\frac{1}{\omega C_s}\right)^2 \tag{8}$$

and Θ can have any value between 0° and 90° depending on the measuring frequency.

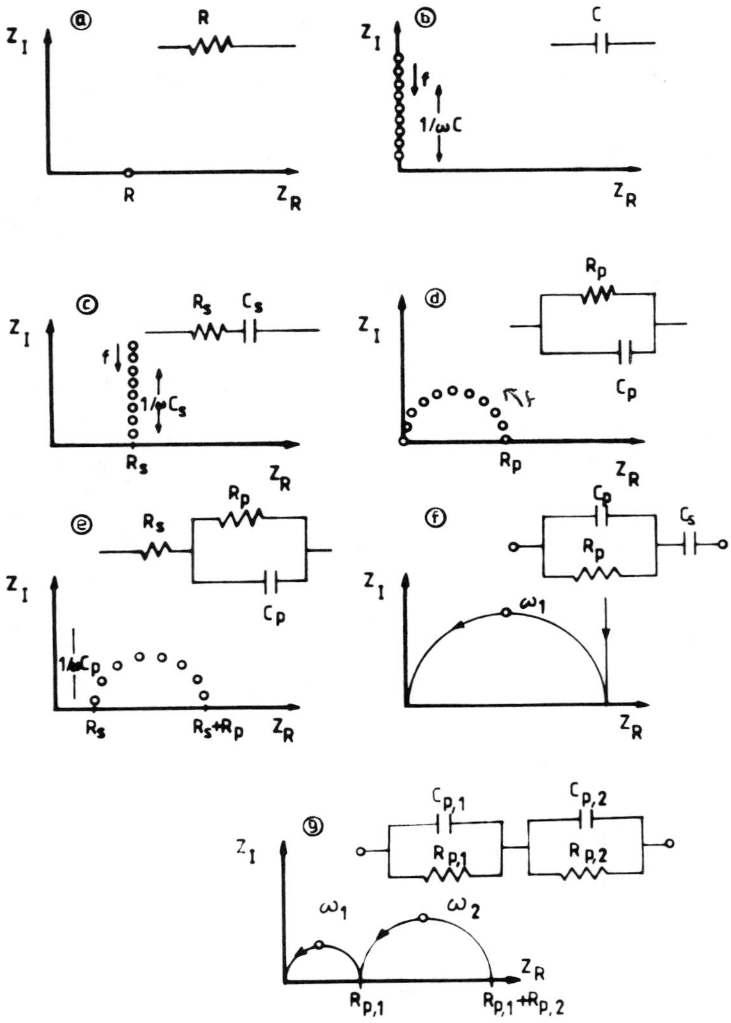

FIGURE 3. Complex plane impedance spectra with their associated equivalent circuits. (From Archer, W. I. and Armstrong, R. D., Spec. Period. Rep. The Chemical Society of London, 1980, 157; Buck, R. P., *Hung. Sci. Instrum.*, 49, 7, 1980. With permission.)

The impedance spectrum of a resistance and capacitance coupled in parallel is shown in Figure 3d. The plot is a semicircle and intersects the real axis at $\omega = \infty$ and $\omega = 0$, respectively. The maximum occurs at $Z_R = R_p/2$, where $Z_I = R_p/2$ and $\omega\tau = 1$. $\tau = R_pC_p$ and is called time constant. The impedance of the circuit is given by

$$Z = \frac{1}{\dfrac{1}{R_p} + i\omega C_p} \tag{9}$$

$$Z = \left(\frac{R_p}{1 + \omega^2C_p^2R_p^2}\right) - \left(\frac{i\omega C_pR_p^2}{1 + \omega^2C_p^2R_p^2}\right) \tag{10a}$$

or

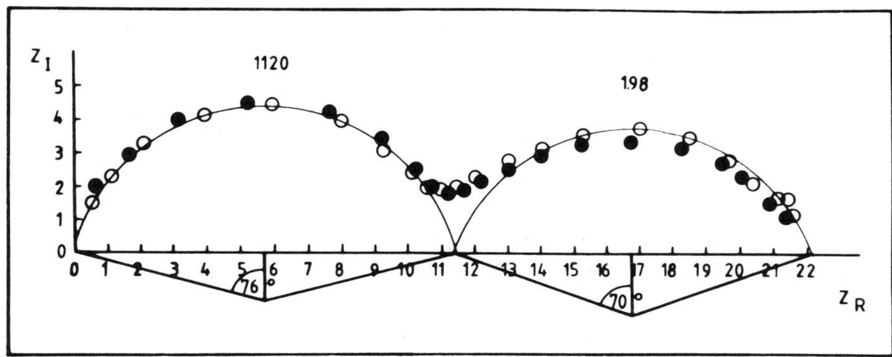

FIGURE 4. Complex impedance plane representation of G.P. (general purpose) glass data at 45°C. Dots are experimental points and circles are theoretical. Z_R and Z_I are given in Mω, ω at peak Z_I is written above each semicircle. (From Buck, R. P., *Hung. Sci. Instrum.*, 49, 7, 1980. With permission.)

$$Z = \frac{R_p}{1 + (\omega\tau)^2} - \frac{i\omega C_p R_p^2}{1 + (\omega\tau)^2} \tag{10b}$$

By comparing Equations 7 and 10a, R_s and C_s can be expressed in terms of R_p and C_p and vice versa:

$$R_s = R_p\left(\frac{1}{1 + \omega^2 C_p^2 R_p^2}\right) \tag{11}$$

$$C_s = C_p\left(\frac{1 + \omega^2 C_p^2 R_p^2}{\omega^2 C_p^2 R_p^2}\right) \tag{12}$$

or

$$R_p = R_s\left(\frac{1 + \omega^2 C_s^2 R_s^2}{\omega^2 C_s^2 R_s^2}\right) \tag{13}$$

$$C_p = C_s\left(\frac{1}{1 + \omega^2 C_s^2 R_s^2}\right) \tag{14}$$

Three other simple circuits and the corresponding impedance plane plots are shown in Figure 3e to g. From the above it follows that one can calculate from the impedance spectrum the components of an equivalent circuit.[9,12,29]

When studying real systems, it was found that the semicircles on the impedance plot are often distorted for three main reasons as reported by Buck:[9]

1. The ratio of the time constants of the parallel RC circuits connected in series is smaller than 10. In such cases, the respective semicircles are overlapped.
2. The equivalent circuit consists of a great number of parallel RC elements in series and the respective time constants are randomly distributed about a mean value. In this case, the semicircle appears "sunken", i.e., the center lies below the real axis (Figure 4).[30]
3. Many parallel RC elements with equal C and R per unit length can be represented as a transmission line. This behavior is recognized as a Warburg diffusional impedance.[9-12,31,32]

B. Application of the Method for Ion-Selective Electrode Studies

From the impedance plane plots of ion-selective electrodes (ISE), the experimental response function of the system can be interpreted in terms of equivalent circuits based on a theoretical model, by considering the transport of charged species through the phase boundaries and the bulk of the membrane.[9-12] The impedance characteristics of a cell incorporating an ISE depends first of all on the properties of the membrane; however, it is also affected by several experimental parameters such as frequency, temperature, and bathing solution activity. Factors that determine equilibrium and steady-state membrane impedances at room temperature were surveyed by Buck[12] and MacDonald:[27]

1. Chemical homogeneity on a microscopic level
2. Physical uniformity and freedom from cracks, grain boundaries, or other unusual transport pathways
3. Number and kind of charge carriers
4. Mobility of charge carriers
5. Rate of transfer of carriers from electrolyte
6. Rate of adsorption when adsorption-reaction paths are involved
7. Rate of generation and recombination of charge carriers from complexes, ion pairs, and lattice sites
8. Location and types of space charge, adsorbed charge, and surface ionic states

Items 5, 6, and 8 are examples of factors that can influence transport across interfaces.

It is not necessary to assume that the same charged components are responsible for building up the space charge at the phase boundary and for the transport of the electric current within the bulk of the membrane. A good example is the glass electrode in which H^+ ion exchange in the swollen boundary phase is the potential determining reaction, but the electric current is conducted by Na^+ ions in the bulk of the glass membrane.[33]

Similar phenomena may also be encountered for other ISE, e.g., in the case of silver halide and sulfide electrodes,[34,35] as well as neutral carrier electrodes (K^+, Na^+, etc.) incorporating lipophilic salts as membrane additives to reduce membrane resistance.[36-39]

The other aim of the impedance studies is the determination and interpretation of the time constants of the individual processes affecting the electrode response functions. If the model used for the interpretation of the time constants is relevant, a good correlation is expected to be found between these data and those obtained by means of independent methods.[23]

C. Information Obtained from Impedance Plane Plots

It is known that a membrane symmetrically or asymmetrically bathed can be perturbed by periodic current or voltage signal as well as by an activity step. The perturbation is considered to be linear, when the input or output voltage is less than RT/zF volts. With linear perturbations, in this sense the time response of the system can be described by an equation consisting of a DC steady-state term and time-dependent terms also:[11]

$$\Phi = \Phi_{dc} + \Phi_n \sum_n \exp(-n\omega t) \tag{15}$$

Theoretical and experimental studies on homogeneous membrane electrode systems by different transient techniques suggest that four or five individual processes, involving coupled R and C elements, may show up on the impedance plane plot (Figure 5; Table 1).

The relevant time constants in order of increasing magnitudes according to Buck[9,12] are as follows:

$$\tau_B = R_B C_B \,^{40} \tag{16}$$

$$\tau_R = R_R C_R \quad [41] \tag{17}$$

$$\tau_F = R_F C_F \quad [30] \tag{18}$$

$$\tau_A = R_A C_A \quad [42] \tag{19}$$

$$\tau_G = R_G C_G \quad [43] \tag{20}$$

$$\tau_w = R_w C_w \quad [11,28,31,32,40,44] \tag{21}$$

The designations are explained in Table 1. At the same time it must be emphasized that with the transient techniques used for studying ISE membranes, the experimental response function of the system cannot always be separated into four or five exponential functions.[9-12,20]

The shortest time constant in the Mc/sec region is related to the dielectric relaxation in liquids and defect motion in solids. This time constant does not appear to be important for ISE.

The next, longer time constant (τ_B) is that due to charging of the external space charge regions coupled to AC bulk electrode resistance. The membrane bulk resistance (R_B) depends on the concentration, mobility, and distribution of the charged particles. The relevant time constant is independent of the thickness and surface area of the membrane. It ranges from 0.3 msec for an AgCl single crystal membrane[45] to 100 to 200 msec for hydrogen ion-selective glass electrodes.[30]

The third time constant, τ_R (Equation 17), which may be observed can be related to the surface resistance and the relaxed double diffuse capacitance coupled in series at each interface. This time constant can be identified only if the surface resistance is in the same order of magnitude as the bulk membrane resistance. Accordingly, when surface processes are rapid and reversible (i.e., the surface resistance is small, which is consistent with a high exchange current density), these processes do not show up on the impedence plane plot and this time constant cannot be determined. However, if the membrane surface is covered by resistive surface layers, the exchange current density may be controlled by the diffusion of ions through this high resistance region and a new time constant, τ_F, can be determined (Equation 18). The resulting τ_F can be distinguished from τ_R by, e.g., etching the surface.[30]

The slow time responses of glass electrodes with protonated and hydrolyzed surface layers and of LaF_3-based fluoride electrode (in the range of OH^- interference) have been attributed to high resistance surface layers. The measured time constants were between 7 to 30 sec.[9,10,20,30]

The longest membrane-controlled time constant, τ_w (Equation 21) is attributed to concentration polarization of the charge carriers within the membrane. This is the Warburg finite-diffusion process resulting in a new steady-state concentration profile. Typical examples for Warburg behavior are liquid, mobile site membranes.[9-12] In the low frequency region, only mobile ions carry current. Accordingly, the Warburg resistance (R_w) corresponds to the DC resistance, while the capacitance (a diffusional pseudocapacitance) is connected to the charge separation due to transport.

It is rather fortunate that in the case of thick ion-selective membranes, the semicircles and the corresponding time constant of the individual processes are often separated. From the resistance, capacitance, and time constant values of different processes, information can be concluded for the nature of charge carriers; thermodynamic properties of complexes and ion pairs; permittivity of membranes; activation parameters of transport; surface rates; and geometric capacitance. Consequently, several important features of glass-,[30,44,46-48] AgCl-,[45,49,50] AgBr-,[50,51] Ag_2S-,[50,52] LaF_3-,[50,53] Aliquat® nitrate-,[54,55] and valinomycin[56-61]-based electrodes had been interpreted with the help of complex impedance methods.

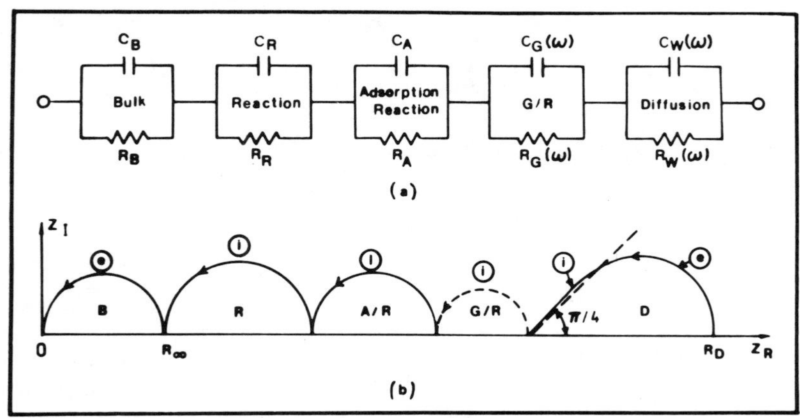

FIGURE 5. (a) Equivalent circuit following from recent models.[27] (b) possible Z (ω) complex plane plot consistent with the circuit in (a). (From Buck, R. P., *Hung. Sci. Instrum.*, 49, 7, 1980. With permission.)

Table 1
PROCESS IDENTIFICATION

τ	R	C
B: Bulk, geometric	R_B = R, resistance for uniform carrier concentration; all charged membrane species carry current, regardless of blocking at interfaces	$C_B = C_g$, geometric capacitance of both electrolyte space charge regions at outer surfaces of membrane (pF/cm^2)
R: Reaction, surface activation overpotential	R_R = surface resistance equivalent to Buttler-Volmer-Erdey-Gruz activation overpotential required to move an ion across the phase boundary	$C_R = C_{d.1.}$ double diffuse layer at an interface (μF/cm^2)
F: Film, resistive surface layer	R_R = true surface region resistance from film such as SiO_2 (or hydrate) on glass	C_F — same as C_R
A: Adsorption, reaction, coupled cases, mainly observed for metals	R_A = surface resistance equivalent to adsorption for systems so irreversible that dissolved species do not react, but specifically adsorbed species have finite, potential-dependent rates	$C_A = C_{d,1}$ doubly diffuse capacitance modified by specific adsorption of charged and uncharged species
G: Generation of charge carriers, recombination of carriers	R_G = modification of R_W by internal replacement or removal of charge carriers from lattice sites, ion pairs, or complexes	C_G = pseudocapacitance (like C_W) because of natural charge separation in concentration profiles of species with different mobilities
W: Warburg diffusion of charge carriers and complex species	R_W = resistance near inner membrane surface occurring by addition of charge carriers from electrolyte at one interface, and loss of charge carriers from opposite interface during half-circle of current flow	C_W = pseudocapacitance in mixtures of ions moving at different velocities because of different mobilities

From Buck, R. P., *Hung. Sci. Instrum.*, 49, 7, 1980. With permission.

III. POLARIZATION STUDIES OF ION-SELECTIVE ELECTRODES

To elucidate electrochemical reaction mechanisms in principle, the following are considered:[20]

1. The sequence of partial reactions contributing to the overall electrode reaction
2. The type of reaction rate control (the type of overvoltage)
3. The reaction orders

Valuable information about partial reactions is obtained especially from the dependence of the total current density j on electrode potential ϵ and concentrations. The total current density j is the sum of the positive anodic partial current density j_a (>0) and the negative cathodic partial current density j_c (<0):

$$j = j_a + j_c \tag{22}$$

At equilibrium potential ϵ_e, the anodic and cathodic current densities compensate for each other, so that no external current flows and no macroscopic reaction takes place:

$$j_a = j_c = j_o \tag{23}$$

j_o is called the exchange current density. It is a measure of the rate of equilibrium potential formation and sensitivity to interference.

Although the exchange current density is known to be one of the most important quantities in electrochemical kinetics, only a few papers deal with its determination in the case of ISE.[13,14,62-66] This appears to follow from the experimental and interpretative difficulties. These problems were studied and surveyed by Cammann:[66]

1. Unlike metal electrodes, the composition of the membrane surface may be altered by the current flow, thus resulting in the change in the chemical potential in that area.
2. Diffusion processes may also occur in the membrane phase, thereby influencing the effective potential difference across the interface which controls kinetic processes.
3. In case of high ohmic membrane bulk resistance, a corresponding ohmic drop voltage develops which is difficult to separate from the small overvoltage caused by low charge transfer resistance (high exchange current density).

Following from these, all values determined for exchange current densities should be regarded as apparent ones (points 1 and 2), loaded with a relatively high measuring error (point 3).[62-66]

A. Methods for Determining Exchange Current Densities

The electrochemical kinetic studies have the advantage to collect information about the details of reactions before the system has reached an equilibrium state. Accordingly, the aim of kinetic studies with ISE is to show the fundamental importance of time-dependent processes on the potentiometric behavior of ISE. This would offer a possibility to elucidate the working mechanism of all types of ISE in the same manner. By considering kinetic data, one may explain some phenomena of ISE which cannot be interpreted on a solely thermodynamic basis, such as sub-Nernstian slope and nonmonotonic transient potentials.[13,14,34,65-67]

In the case of charge transfer polarization, the relationship between potential and total current density, j, is given by Buttler-Erdey-Gruz-Volmer equation:[68,69]

$$j = j_o\left[\exp\left(\frac{\alpha zF}{RT}\eta\right) - \exp\left(-\frac{(1-\alpha)}{RT}zF\eta\right)\right] \tag{24}$$

where j is the current at overpotential η; j_o is the exchange current density at $\eta = \epsilon - \epsilon_e = 0$; and α is the transfer coefficient. The exchange current density equals the partial current densities that compensate for each other at equilibrium potential (ϵ_e):

$$\begin{aligned} j_o &= zFk_a \cdot c_r \cdot \exp\left(\frac{\alpha F}{RT}\epsilon_e\right) \\ &= zFk_c c_o \cdot \exp\left(-\frac{(1-\alpha)\,zF}{RT}\epsilon_e\right) \end{aligned} \tag{25}$$

where c_r and c_o are the concentrations of reduced and oxidized species, respectively; k_a and k_c are constants; and z, F, R, T have their usual meaning.

There are basically two methods for the determination of the exchange current density from Equation 24. At small overpotentials ($|\eta| \ll RT/zF$), the exponential terms of Equation 24 can be developed in series, the terms at higher power can be neglected, and the transfer coefficient drops out:

$$j = j_o \cdot \frac{zF}{RT}\eta \tag{26}$$

The proportionality factor between the overvoltage η and the current density j corresponds to an electrical resistance designated as charge transfer or polarization resistance:

$$R_t = \left(\frac{d\eta}{dj}\right)_{j=o} \tag{27}$$

Accordingly, the exchange current density can be given as

$$j_o = \frac{RT}{zF} \cdot \frac{1}{R_t} \tag{28}$$

Alternatively, j_o can be determined at high overpotentials ($|\eta| \gg RT/zF$) where the first or second term in Equation 24 can be neglected depending on the sign of the current, and a linear relationship exists between charge transfer overvoltage and the logarithm of current density:

$$\eta = -\frac{RT}{\alpha zF}\ln j_o + \frac{RT}{\alpha zF}\ln j \tag{29a}$$

or

$$\eta = \frac{RT}{(1-\alpha)\,zF}\ln j_o - \frac{RT}{(1-\alpha)\,zF}\ln|j| \tag{29b}$$

and j_o is obtained by extrapolation to the equilibrium potential, $\eta = 0$.

These methods are valid if the electrode reactions are exclusively controlled by charge transfer processes (concentration polarization is not involved). Accordingly, if only charge

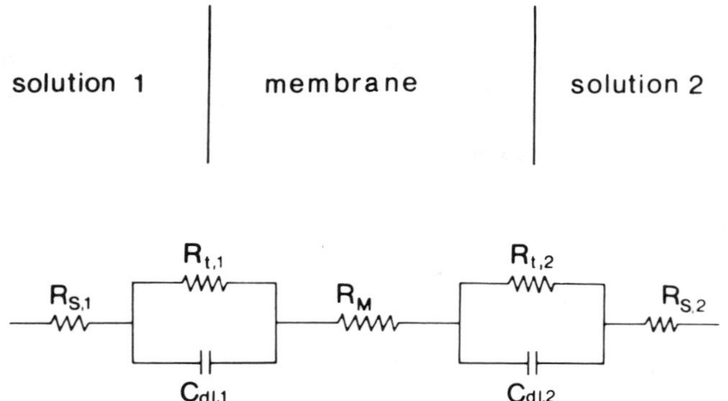

FIGURE 6. Simplified equivalent circuit for membrane electrodes: R_{S1};R_{S2} are solution resistances; R_M is the membrane resistance; $R_{t,1}$;$R_{t,2}$ are charge transfer resistances at membrane/solution interfaces; and $C_{dl,1}$;$C_{dl,2}$ are double layer capacitances at membrane solution interfaces. (From Cammann, K. and Rechnitz, G. A., *Anal. Chem.*, 48, 856, 1976. With permission.)

transfer overvoltage is assumed, the equivalent circuit of an electrode symmetrically bathed is shown on Figure 6. It is apparent that reliable information on the charge transfer resistance can be obtained only if

$$R_t \geqslant R_M/100$$

assuming 1% measuring accuracy. Unfortunately, this condition is not fulfilled in the case of most ISE. Attempts to decrease membrane resistance R_M by decreasing the membrane thickness failed as the decrease in membrane thickness is limited from practical points of view.

Another way of reducing membrane resistances is to incorporate lipophilic salts into liquid membranes.[36-39] However, the amount of added lipophilic salts is strictly limited, since its excess, compared to that predicted by theory, leads to drastic changes in membrane selectivity.[37,70-72]

The charge transfer overpotential of ISE was almost exclusively studied by the so-called galvanostatic current step method.[2] In the galvanostatic current step method (Figure 7), the working electrode at equilibrium is polarized with a small amplitude, rectangular current signal and the resulting potential vs. time function is recorded and analyzed (Figure 8).

The latter is described by Equation 30 if (1) the membrane is a pure ionic conductor; (2) only charge transfer overvoltage is encountered; (3) the current density is low enough to be in the linear range of the current voltage curve; and (4) ohmic drop on the solution layer between the membrane and the reference electrode is relatively small, and thus can be neglected:[62]

$$\Phi = jR_M + j\left[R_t - R_t \exp\left(-\frac{t}{R_t C_{dl}}\right)\right] \tag{30}$$

The ohmic potential drop determined by the membrane resistance (R_M) can be obtained from Equation 30 at $t = 0$:

$$\Phi(t = 0) = jR_M \tag{31}$$

while at $t \rightarrow \infty$:

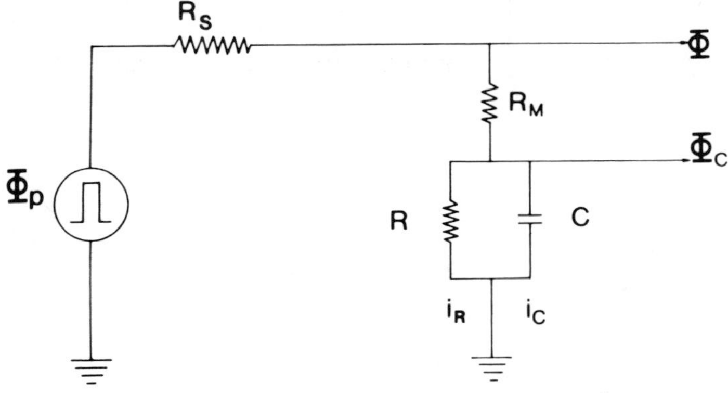

FIGURE 7. Schematic diagram related to the method of galvanostatic current step. R_M is the membrane resistance; $R = 2R_t$ and $C = C_{dl}/2$ (if $R_{t,1} = R_{t,2}$ and $C_{dl,1} = C_{dl,2}$, compare Figure 6). Further on, $R_S \gg R_M + R$; accordingly, $I = \text{const} \approx \Phi_p/R_s$.

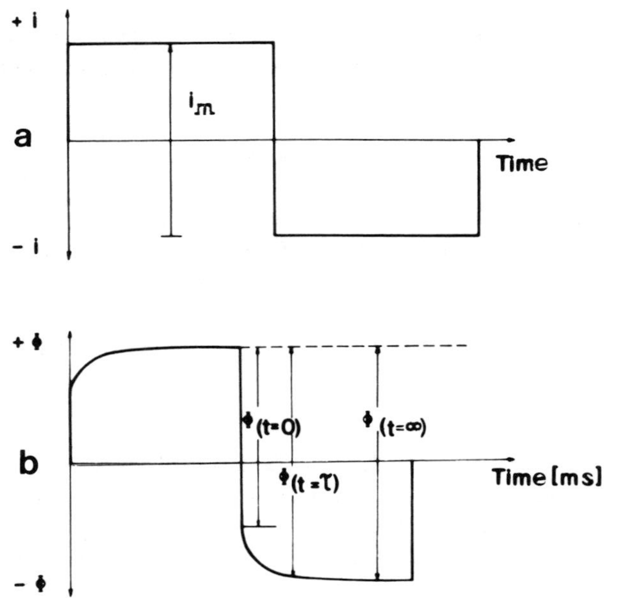

FIGURE 8. Input (a) and output (b) signals employing galvanostatic perturbation (Figure 7).

$$\Phi(t \rightarrow \infty) = j(R_M + R_t) \tag{32}$$

R_M and R_t are determined from Equations 30 and 31, while the capacitance of the double layer is evaluated from the time constant τ ($\tau = RC$).

At $t = \tau$, Equation 30 becomes (Figure 8)

$$\Phi(t = \tau) = j(R_M + R_t + R_t e^{-1}) = j(R_M + 0.632\ R_t) \tag{33}$$

However, the double layer capacitance can also be determined from the slope of the Φ vs. t function:

$$\frac{d\Phi}{dt} = \frac{j}{C_{dl}} \exp\left(-\frac{t}{R_t C_{dl}}\right) \tag{34}$$

since at t ~ 0, the exponential term can be simplified:

$$e^{-x} \approx (1-x) \approx 1$$

and thus

$$C_{dl} = j\left(\frac{dt}{d\Phi}\right)_{t\to 0} \tag{35}$$

The measuring time of the galvanostatic current step method should be chosen in such a way that the diffusion overpotential could be neglected. To calculate this time value, the Sand equation can be applied:[2,64,73]

$$\eta_D = \frac{RT}{zF} \ln\left(1 \pm \sqrt{\frac{t}{T}}\right) \tag{36}$$

where

$$T = \frac{\pi}{4} D\left(\frac{zFc}{j}\right)^2 \tag{37}$$

and where + is the sign for anodic current; − is the sign for cathodic current; D is the diffusion coefficient; and T is the transition time. The longer the measuring time is chosen, the less the diffusion overvoltage can be neglected.

As the transition time is a quadratic function of concentration, the measuring time to be applied decreases quadratically also with decreasing concentration. The method of Delahay and Berzins[74-76] permits the correction of the diffusion overpotential in the galvanostatic current step technique. In this method, the voltage-time relationship, after a short time delay, is given by

$$\eta = \frac{-2j\ RT\ t^{1/2}}{\pi^{1/2} z^2 F^2 c_o^o\ D^{1/2}} + \frac{jRT}{2F}\left[\frac{RTC}{z^3F^3(c_o^o)^2 D} - \frac{1}{j_o}\right] \tag{38}$$

where t is the time; c_o^o is the bulk concentration; C is the double layer capacitance; D is the diffusion coefficient of the ion involved; and z is the charge of the ion.

According to Equation 38, the η vs. $t^{1/2}$ function is linear and the diffusion coefficient can be determined from its slope value:

$$\frac{d\eta}{d(t^{1/2})} = \frac{-2\ RT\ j}{\pi^{1/2} z^2 F^2 c_o^o D^{1/2}} \tag{39}$$

If the double layer capacitance is known, j_o can be evaluated by extrapolation to t = o:

$$\eta = \frac{jRT}{zF}\left[\frac{RTC}{z^3F^3(c_o^o)^2 D} - \frac{1}{j_o}\right] \tag{40}$$

The precision of the determination of exchange current densities is primarily affected by the iR_M ohmic potential drop.

Since the exchange current density is concentration dependent, the ratio R_t/R_M can be increased by reducing the concentration of the potential determining ion (see Chapter 3, Section II.A). However, this approach includes a basic difficulty: by decreasing the concentration of the appropriate potential determining ion, side reactions related to the other constituents of the solution may show up more explicitly. The extrapolation of the log concentration vs. log R_t function to 1 *M* primary ion concentration yields the apparent exchange current density for this ion, provided that all other resistances in the cell are held constant.[65]

B. A Mixed Potential Ion-Selective Electrode Theory

Ion-selective electrodes (ISE) are based on such membranes which hinder but do not exclude the transport processes between the contacting phases. The material transport may include simple ions, neutral or charged complexes, electrons or mobile holes, and it is of special analytical interest because it can generate potential differences which depend on the activity of ions or molecules present in the outer phases in contact with the membrane. A sufficiently thick ion-selective membrane contacted with solutions on both sides can be considered a homogeneous phase with two boundary layers.

The potential determining processes taking place at the interfaces (including the material transport through this interface) and within the bulk of the membrane can be separated. Buck[22] reviewed the various potential generating processes encountered on blocked and nonblocked interfaces.

The theories related to the interpretation of the potential response mechanism of ISE can be divided into two groups. The first one is the so-called three-segmented potential model based on the works of Donnan,[77,78] Theorell,[79] Meyer and Sievers,[80] Helferich,[81] Schlögl,[82,83] Eisenman et al.,[84-86] and, more recently, of Morf et al.[17,87] The second group includes the interfacial potential theories employed first by Nicolsky et al.[88,89] for the interpretation of the pH response of glass electrodes and later applied to ISE by Pungor,[90,91] Buck,[92] and others.

The segmented membrane model is based on the assumption of thermodynamic equilibrium at the interfaces. Deviations from the ideal Nernstian potential response are interpreted by assuming that diffusion potential is generated inside the ion-selective membrane. The diffusion potential can be expressed, e.g., by the Nernst-Planck equation.

The interfacial potential model is restricted to the charge distribution at the phase boundaries. The formation of the space charges is affected by the properties of the homogeneous membrane phase. It is to be noted, however, that the strict definition of phase boundary — i.e., an interface in which given physical properties change discontinuously according to a step function[14] — does not hold practically, because many important physical properties such as density and viscosity vary continuously at the phase boundary of two phases in a microscale.

Another approach for the interpretation of the phase boundary potential is a kinetic one. This was first introduced by Buttler[68] and Erdey-Gruz and Volmer,[69] and was applied more recently by Cammann[13,14,93] for the interpretation of the potential response of ISE. This is based on kinetic phenomena appearing on the interfaces and is called the electrokinetic or mixed potential model.

The number of the charge carrying components crossing the interface at $t = 0$ is defined by the activation energy of the electrochemical reactions and the concentration of the species taking part in the process. As usual, the charge transport at the interface towards the membrane decreases the current flowing in the same direction, while it increases the current flowing in the opposite one. As a result, an equilibrium is established when the two partial currents are equal, and thus no external current can be measured.

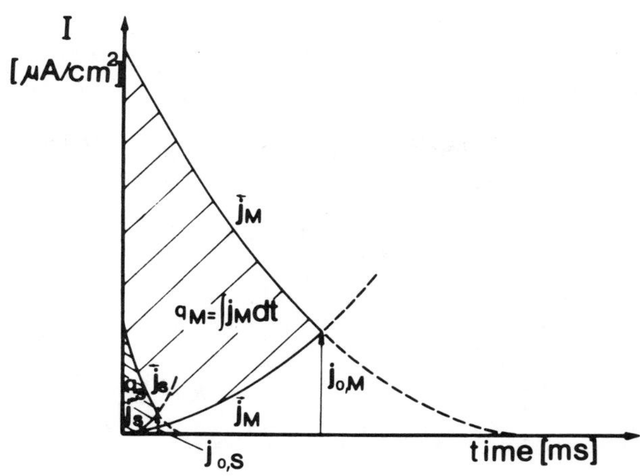

FIGURE 9. Schematic diagram of the directed currents $\vec{j}$ and $\overleftarrow{j}$ across the interface of an ion-selective membrane. M = primary ion; S = interfering ion; q_M,q_s = net charge transported by the primary and interfering ions, respectively; and $j_{o,M}$,$j_{o,S}$ = exchange current density for primary and interfering ions. (From Cammann, K., *Ion-Selective Electrodes*, Pungor, E., Ed., Akadémiai Kiadó, Budapest, 1978, 297. With permission.)

In Figure 9, an idea is presented for visualizing the charge distribution model as a function of time for the primary and interfering ions. It can be seen that the potential determining ion is that which has the greatest contribution to the change of the charge distribution.

From this it follows that the potential corresponding to the fastest electrode reaction acts as an overpotential for the other reactions.[13,14] In spite of this, interfering ions reach an equilibrium condition also. Consequently, the relative amount of charge carriers, transported by the primary and interfering ions across the interface, determines the selectivity of the electrochemical sensor.[13,14,34,65,66]

Since all equations that describe the charging process contain concentration data, it is obvious that the dynamic response characteristics of an ISE are also concentration dependent (see Chapter 3, Section II.A).

With the help of the above kinetic model (Figure 9), the nonmonotonic transient signals measured in the presence of both primary and interfering ions can be interpreted. Following up a fast activity change of interfering ions, the charge transported to the surface of the electrode membrane by the interfering ions can be considerably higher than its equilibrium value, which means that its contribution to the overall potential is much higher than at equilibrium. This is estimated as a potential overshoot, the relaxation of which is controlled by $t^{-1/2}$ function (see Chapter 4, Section III).

If the rate of the kinetics for primary and interfering ions is similar, parallel electrode reactions have to be taken into account. Thus, the equilibrium potentials of the individual processes affect each other and the resulting steady-state potential lies between the equilibrium potential corresponding to that defined by primary and interfering ions separately.

An exact theoretical treatment of this process is difficult even in the case of parallel redox reactions at metallic electrodes.[2,93,94] Nevertheless, a qualitative picture (shown in Figures 10 and 11) can give some insight into processes existing at the face boundary. From this it is obvious that the mixed potential is mainly determined by electrode reaction having the steepest current voltage characteristic (see also Chapter 3, Section II.B). The slope of the current vs. voltage curve in the range of equilibrium potential corresponds to the charge transfer resistance, R_t, which is inversely proportional to the exchange current density (Equation 28).

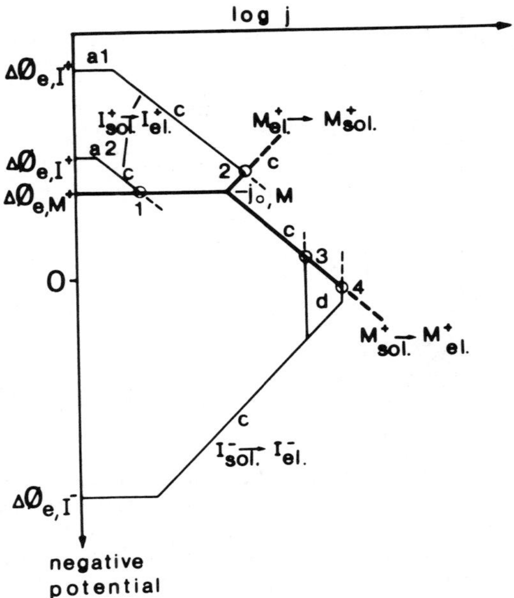

FIGURE 10. Evans diagram with hypothetical log-current-voltage curves. $\Delta\Phi_e$ = equilibrium potential; M^+ = cation to be measured; I^+ = interfering cation; I^- = interfering anion; sol. = solution phase; el. = electrode phase; a_1,a_2 = different activities of the interfering cation (1,2); c = range of kinetic control; d = range of diffusion control; j_o = exchange current density; circles = mixed potential; 1 = ideal Nernst sensor; 2 = interference; and 3,4 = stirring sensitivity. (From Cammann, K., *Ion-Selective Electrodes,* Pungor, E., Ed., Akadémiai Kiadó, Budapest, 1978, 297. With permission.)

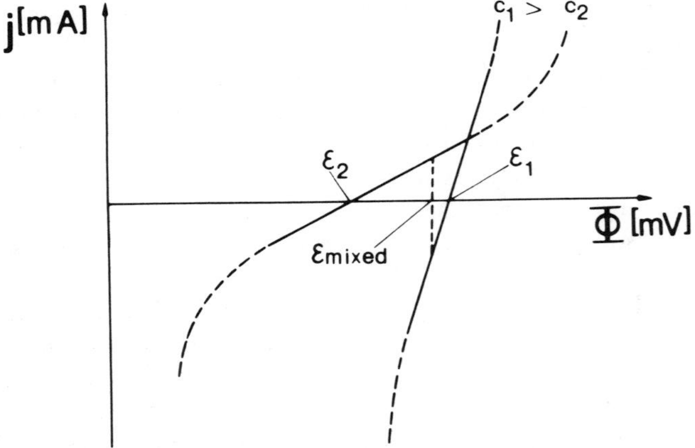

FIGURE 11. Hypothetic current voltage curves for ISE recorded at two different primary ion concentrations.

For a full description of the behavior of an ISE with the mixed potential model, it is not sufficient to know the exchange current densities of the different charge transfer processes; but it is necessary to know the whole current voltage characteristics.

The latest results in the field of electrolysis at the interface of two immiscible electrolyte solutions (ITIES)[95] may offer a possibility to verify the mixed potential model of ISE.

Summing up, it can be concluded that the kinetic model may offer a possibility for the description of the potential response mechanism of ISE without the assumption of internal diffusion potential. The model intends to interpret the ISE response mechanism by superimposed current-voltage curves of parallel electrode processes, the single parameters of which may be determined independently.

Accordingly, the potential differences are created only at phase boundaries, and thus the short response times of thick membranes are no longer inexplicable. However, transfer rates at the phase boundaries are dependent on the composition of the bulk and the surface of the membrane.

IV. STUDIES INDIRECTLY RELATED TO THE DYNAMIC CHARACTERISTICS OF ION-SELECTIVE ELECTRODES

The processes supposed to affect the dynamic response characteristics of ISE were also studied indirectly by nonelectrochemical methods.[15,16,24,25,96-112]

The time dependence of ion-exchange and adsorption processes on both sides of the phase boundary of precipitate-based electrodes were followed by radioactive tracer techniques.[99,103-107] Atomic adsorption spectrophotometry[24,25,100-102] and conductometric methods[24,25] were applied to tail changes in the solution phase as a result of membrane solution interaction. The kinetics of transport and partition processes in liquid membrane were studied in detail with radioactive tracer techniques[105] and IR[110] and FTIR-ATR spectroscopy[111,112] in the membrane phase and with atomic absorption[15,16] and UV spectrophotometry[98,108,109] in the liquid membrane contacting solution phase.

Radiochemical measurements used for the determination of the kinetics of ISE are in some respect advantageous compared to the methods described in Sections I and II. The membrane resistance does not affect the measurements, since polarizing voltage is not applied. In addition, since the radiochemical experiments were carried out in the absence of current, they avoid all the disadvantages caused by current flow, namely, the change in the chemical composition of the membrane phase and the generation of diffusion-controlled interfacial processes.[13,14] However, it must be noted that radiochemical techniques permit the determination of the slowest step of the overall electrode process only. The latter can be

1. Diffusion of ions from the bulk of the solution to the surface of the electrode
2. Interfacial reactions (adsorption, dehydration, chemical reaction, charge transfer, etc.)
3. Transformations of the products of interfacial reactions (crystallization, coupled chemical reaction, hydration, etc.)
4. Transport of the products of the interfacial reactions into the bulk of the solution or into the membrane

A further problem is connected to the radiochemical methods: the transport of charged particles across the phase boundary is not governed only by the electrochemical potential difference as in potentiometric experiments, but can also be affected by the kinetics and equilibrium of isotope-exchange reaction. To the latter, solid-phase diffusion processes can also be coupled, which can complicate the interpretation of the result in the electrochemical sense.

Zimens[106] and Reber[107] have studied the ion-exchange processes on silver bromide both

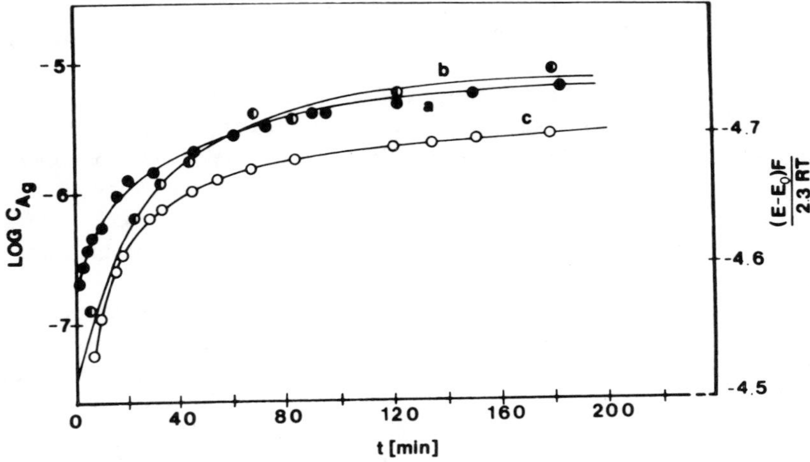

FIGURE 12. Test of dissolution theory for chloride electrode. Measurement made in 10^{-5} *M* chloride solution. T = 25°C; I = 0.1 *M* KNO_3. Curve a: plot of log c_{Ag} vs. t measured by conductometry. Curve b: plot of log c_{Ag} vs. t measured by AAS. Curve c: plot of (E − Eo) F/2.3 RT vs. t. (From Buffle, J. and Parthasarathy, N., *Anal. Chim. Acta,* 93, 111, 1977. With permission.)

in crystalline form and in finely dispersed state on photographic plates using radioactive silver and bromide as tracers. From their work, it can be concluded that practically only silver ions take part in the ion-exchange process; moreover, at a relatively short measuring time, the rate of the ion-exchange process is dominant. However, at a long measuring time, the slowest process is the diffusion of the radioactive isotopes in the solid phase.

Koebel[62] evaluated the exchange current density (j_o = 1.6×10^{-3} A cm^{-2}) and the transfer coefficient (α = 0.55) by extrapolation from the exchange rates determined in 1 *M* silver ion solution by Ziemens[106] and Reber[107] under entirely different experimental conditions. Naturally, the determination of the exchange current density by radiochemical method is only possible if the rate of the exchange reaction is slower or at least commensurable to the rate of diffusion in the solid phase. Peschanski[113] examined the silver ion exchange on β-Ag_2S crystals with radiotracer technique, and found that the diffusion of the silver ion in the solid phase controls the reaction rate. This result is in agreement with the exchange current density data determined with Ag_2S-based silver ISE ($j_o > 1$ A cm^{-2}).[13,65]

The correlation studies between the dynamic response of ISE and the kinetics of dissolution of the active material of solid-state ion-selective membranes contributed greatly to the interpretation of the transient functions of ISE. The amount of silver dissolved in 1×10^{-5} *M* chloride solution contacting a AgCl-based electrode was measured by both conductometry and atomic absorption spectrophotometry.[24,25] The logarithm of the silver concentration determined experimentally and calculated on the basis of electrode potential values is plotted as the function of time and is shown in Figure 12.

From this figure it is clear that the time required to attain the dissolution or electrochemical equilibrium is of the same order of magnitude. Thus, in very dilute solutions, the dissolution kinetic of the electrode membrane plays a central role in the overall electrode reaction.

The kinetic behavior of a series of ligands in the transfer from plasticized PVC membranes into a stirred aqueous phase was investigated and compared with a theoretical model.[96,97] The lipophilicity[114] of the ligand was found to determine the mechanism of transport processes.[96,97] The rate of transfer of the ligand is controlled by the diffusion in the membrane with ligands of low lipophilicity, and by the exchange reaction at the phase boundary or the diffusion through the unstirred (adhering) diffusion layer with ligands of high lipophilicity.

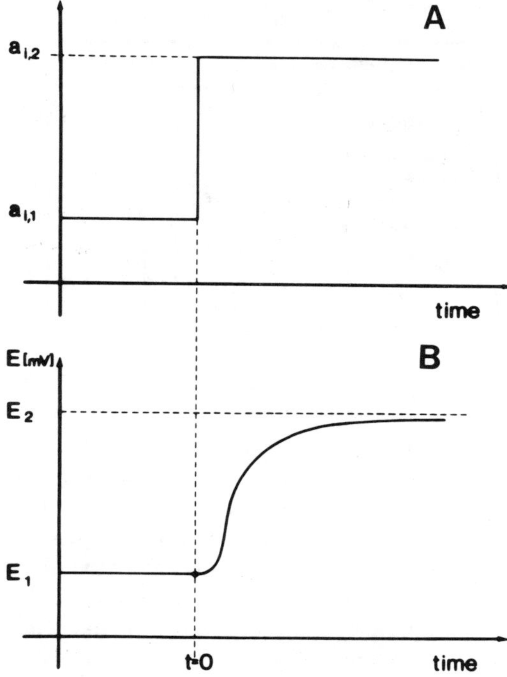

FIGURE 13. The principle of the activity step method: (A) the activity step applied; (B) the cell voltage (E) vs. time function recorded.

These investigations provide a good estimate for the lifetime of neutral carrier-based ion-selective liquid membrane electrodes; further on, they may contribute to a better understanding of the dynamic characteristics of these type of sensors. This can lead to finding a compromise in designing a new sensor of short response and long life time.

V. THE ACTIVITY STEP METHOD

In the use of ISE, it is often assumed that the rate of attaining equilibrium electrode potential is rapid compared to the rate of the studied process. This assumption, however, cannot always be accepted in the studies of the kinetics of fast reactions, in analyzers monitoring high speed processes, or in fast serial analyzers, etc. The application of hydrogen ion-selective glass electrodes for monitoring of rapid physiological processes,[33,115] for the study of the kinetics of chemical reactions,[116-118] and for the control of industrial processes[119,120] raised the importance of the examinations of the dynamic properties of ISE with the activity step method.[121-124] In the late 1960s, similar studies were performed with other types of ISE,[4,5,125-127] mainly with the aim of providing dynamic data for practical purposes.

A. The Principle of the Method

In the activity step method, the perturbing signal is an activity step accomplished in the solution phase contacting one side of the membrane (Figure 13), and the resultant cell potential is recorded as a function of time (Figure 13).

When applying the activity step method, the measured overall dynamic response signal (i.e., the transient function) is affected by the properties of ISE, but also by the design of the electrochemical cell, the measuring technique applied to realize the activity step, the

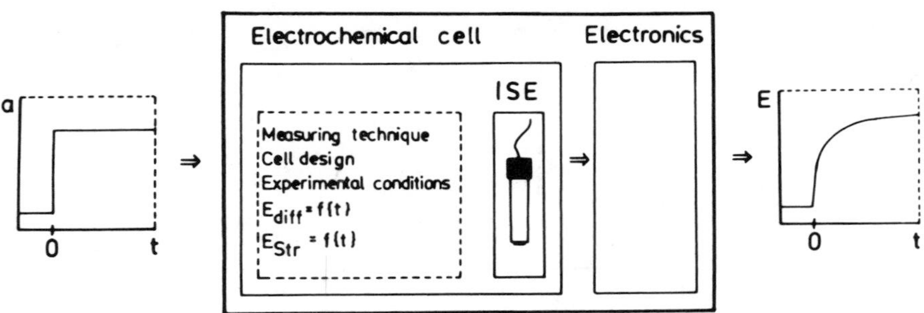

FIGURE 14. Schematic diagram of the measuring system used with the activity step method.

experimental conditions selected, the time dependence of other sources of potential in the cell, (e.g., diffusion and streaming potentials), and, finally, the electronics used to record the transient signals (Figure 14).[128] All these factors simultaneously define the transient function of the overall measuring system (electrochemical cell and the measuring electronics).

In most analytical applications of ISE, the transient function of the measuring system is determined primarily by the properties of the electrochemical cell with the exception of electrodes of extremely high internal resistances (e.g., microelectrodes).[129-132] Among the parameters affecting the transient function of the electrochemical cell, in addition to the properties of the ion-selective membranes, the ideality of the activity step at the electrode surface is of utmost importance.

The shape of the activity step is influenced mainly by the method used to realize the activity step, but can also be affected by the experimental conditions, e.g., flowrate, viscosity of the sample solution, and surface wetting. The so-called immersion or dipping techniques[7,25,123,133-136] and the injection method[7,25,125,137-140] are most often used as measuring techniques in addition to their versions adapted to continuous analyzers.[141-143] Several apparatuses designed for recording and studying transient functions of ISE were also reported.[4,6,8,115,144-149]

B. Experimental Techniques

1. The Dipping (Immersion) Method[7,25,123,133-136]

The indicator electrode is conditioned in a solution of activity $a_{i,1}$, then after careful wiping or washing, it is immersed into the vigorously stirred sample solution of activity $a_{i,2}$ at the time $t = o$.

The timing starts at the instant of immersion. In order to avoid undesirable perturbations on the response function induced by wiping, the adhering droplets of the conditioning solution can be removed by shaking the electrode during transfer.[136] A serious drawback of this method is due to the fact that the electric circuit is disconnected when the electrode is taken out of the solution. The electrostatic charge accumulated discharges on the capacitance of the double layer and/or on the capacitance parallel to the input of the measuring amplifier.[140,144] Different flow-through techniques,[141-144] e.g., the injection method,[145] avoid these problems and improve the reproducibility.

2. Injection Method[7,25,125,137-140]

A concentrated solution of primary ion in a small volume is injected by means of a syringe into the rapidly stirred test solution in which the electrode had been previously conditioned. The timing is started at the instant of injection of the concentrated sample. Reproducible response time data can only be achieved under exactly controlled hydrodynamic conditions,

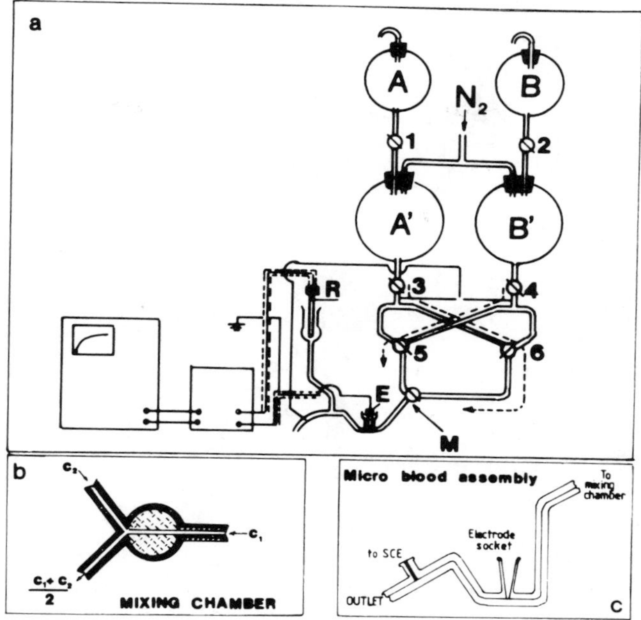

FIGURE 15. Measuring set-up of Rechnitz and Kugler. (a) Schematic diagram of rapid mixing, continuous flow apparatus. E = cation-sensitive glass electrode; M = mixing chamber; and R = reference electrode. (b) Details of mixing chamber. (c) Schematic diagram of flow assembly showing electrode positions. (From Rechnitz, G. A. and Kugler, G. C., *Anal. Chem.*, 39, 1982, 1967. With permission.)

i.e., using the same container, identical solution and injected volume, stirrer geometry, stirring rate, electrode surface area, and depth and angle of immersion of the measuring electrode.[128]

The fact that the electrical circuit is not interrupted by the change of solution provides the advantage of the injection method.[140,144] However, it is a severe drawback of the technique that, in practice, identical mixing conditions cannot be ensured when the effect of an activity decrease is studied.

It must also be taken into consideration that the time of mixing of two solutions of different activity is often not negligible compared to the kinetic of the processes under investigation.

3. Special Techniques

Since the activity steps produced with any of the techniques discussed above were far from being ideal (i.e., with infinite slope at the leading edge) at the surface of the ISE, several special cells and instruments were constructed for dynamic repsonse studies. Among these, the devices developed by Disteche and Dubuisson,[115] Rechnitz and Kugler[144] (Figure 15), Tóth et al.[4-6] and Lindner[8] (Figure 16), Mertens et al.[141] (Figure 17), Denks and Neeb[146] (Figure 18), and Hayano et al.[147] are worth mentioning.

Fast activity changes can be produced either by fast mixing of solutions of different activities (injection method) or by instantaneous replacement of one solution by another (dipping method). At the first period of response time studies on glass[115] and precipitate-based electrodes,[133] the fast displacement of solutions of different activities provided excellent results. In the measuring setup of Disteche and Dubuisson,[115] and much later in that of Nagy and Fjeldly,[148] the ISE and reference electrode faced each other with a separation gap of about 1 mm, and samples of test solution were injected into this gap parallel to the measuring surface by means of a jet. According to the authors, the efficiency of the washing mainly

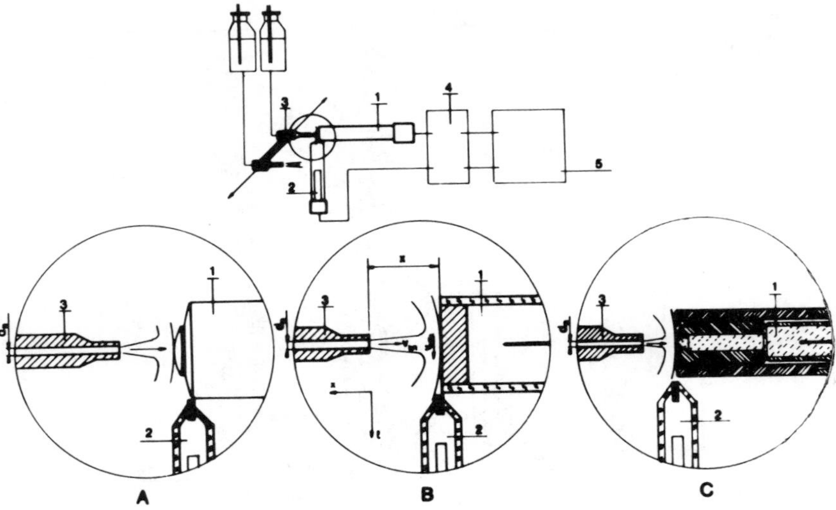

FIGURE 16. Wall-jet cell arrangement for the study of the transient functions of ISE. The flow pattern on the surface of the different ISE: (A) commercial precipitate-based ISE (Radelkis, Hungary); (B) laboratory-made precipitate-based electrode; (C) commercial liquid-membrane electrode (Philips IS-560). 1 = indicator electrode; 2 = reference electrode; 3 = jet; 4 = differential amplifier (Keithly Type 604); and 5 = storage oscilloscope (Philips PM 3251) or x-y recorder (Bryans 2600 A3). x = the distance between the jet and the surface of the ion-selective membrane; d_n = the jet diameter; v_x = the linear flow rate perpendicular to the electrode surface; and v_{str} = the linear flow rate parallel with the electrode surface.

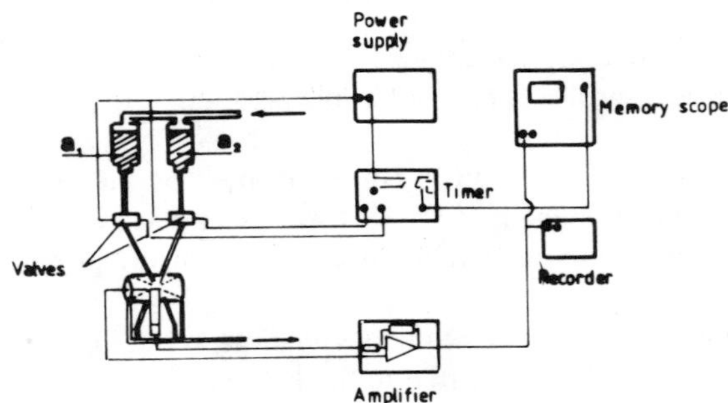

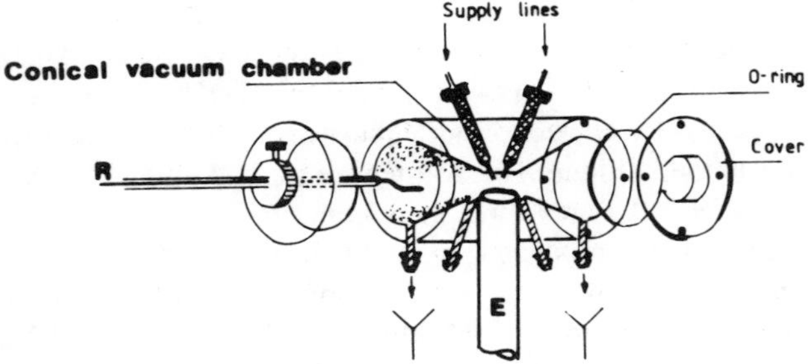

FIGURE 17. Experimental setup for the fast flow injection technique and the perspex conical vacuum chamber used by Mertens et al. (From Mertens, J., Van den Winkel, P., and Massart, D. L., *Anal. Chem.*, 48, 272, 1976. With permission.)

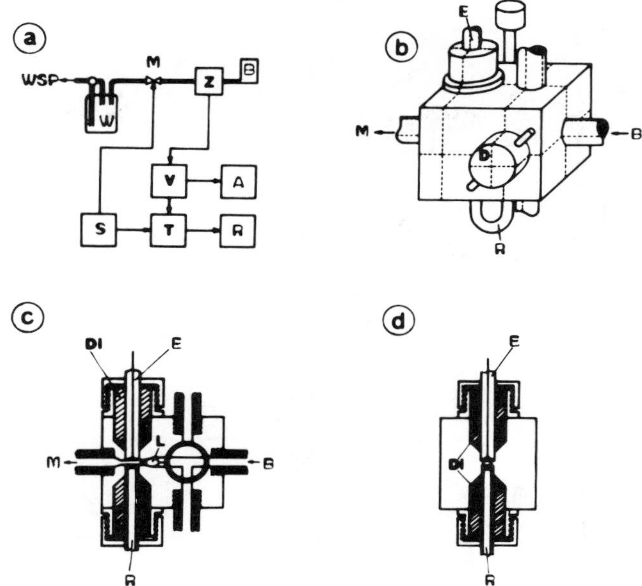

FIGURE 18. Experimental setup and measuring cell used by Denks and Neeb. (a) WSP = vacuum pump; M = magnetic valve; Z = measuring cell; V = differential amplifier; T = transient recorder; R = recorder; and S = controller unit. (b) E = indicator electrode; R = salt bridge; D = three-port tap; and B-M = flow-through channel. (c,d) R = reference electrode; D_i = packing; and L = baffle. (From Denks, A. and Neeb, R., *Fresenius' Z. Anal. Chem.*, 285, 233, 1977. With permission.)

depends on the respective orientation of the jet and the electrodes. Under optimal condition, the remaining solution droplet could be removed in 6 msec. The potential time response of pH glass electrodes followed an exponential type of equation.

Referring to the theoretical work of Hill,[150] Disteche and Dubuisson[115] pointed out first that diffusion processes through a stagnant solution film may complicate the overall response function as a consequence of bad positioning of the jet.

Tóth et al.[4-6] designed a switched wall-jet cell and a connecting measuring instrument in order to ensure a relatively simple and rapid change of two solutions of different activities (Figure 16). The operation principle of the equipment is close to that of the dipping method as far as, instead of changing the position of an electrode, that of the test solutions is changed in a convenient way, by means of two mobile jets placed perpendicular to the electrode surface (Figure 16). A more recent version is shown in Figure 19.[149] During measurement, but also during solution change, a thin electrolyte film ensures the galvanic contact between the indicator electrode and the reference electrode; thus, the disadvantages of the dipping method are avoided.

However, for ensuring a more reliable contact between the measuring and reference electrode, especially at high flow rate and small electrode surface, the cell has recently been developed further by Lindner et al. (Figure 19).[149] The measuring system is connected to a data acquisition system, too.

The measuring systems of Rangarajan and Rechnitz[145] as well as those of Rechnitz and Kugler[144] can be considered improved versions of those employed with the injection technique.

The concentration step required for the response time measurement is produced by the injection of a more diluted or a more concentrated solution plug into the solution stream

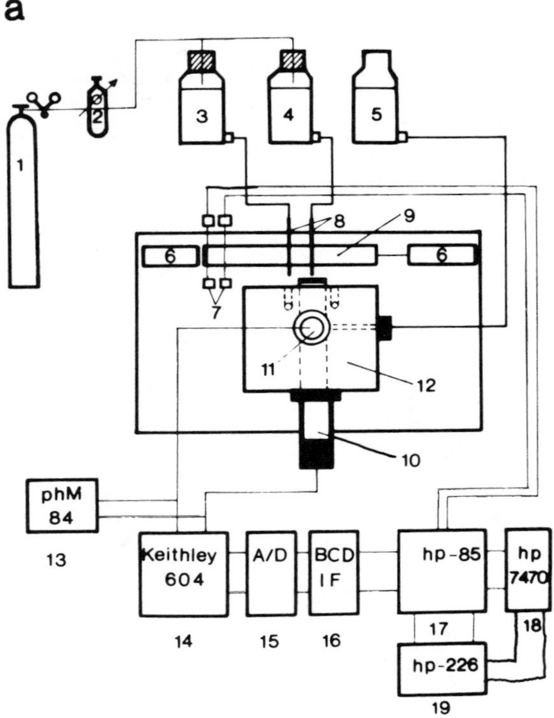

FIGURE 19. Measuring setup (a) and wall-jet cell arrangement (b) used for response time studies of commercially available macroelectrodes. (1) Nitrogen bomb; (2) high precision reduction valve; (3,4) sample solutions of different concentrations; (5) salt bridge electrolyte; (6) pneumatic pistons; (7) optical detectors; (8) jets; (9) mobile clamping arrangement for the jets; (10) ISE; (11) reference electrode; (12) electrochemical block; (13) pH meter; (14) impedance transformator (Keithley 604); (15) AD converter; (16) HP 82941 A BCD IF; (17) HP-85 computer; (18) HP7170 digital plotter; and (19) HP226 computer.

flowing at a high flow rate. The injected solution is introduced in the flow-through detector cell through a mixing chamber, specially designed by Rechnitz and Kugler[144] on the basis of the work of Meier and Schwarzenbach.[117] A complete mixing was attained in $t < 5$ msec at a very high flow rate (7.4 m/sec). The equipment was used mainly for the study of the transient functions of microglass electrodes.

Denks and Neeb[146] (Figure 18) and later on Hayano et al.[147] developed a measuring setup in which the activity step is accomplished without the use of a stirring unit.

High linear flow rate (3 m/sec,[4-6] 6.8 m/sec,[146] 7.4 m/sec,[144] 11.8 m/sec[147]) is used in all of the equipments discussed so far in order to ensure an ideal activity step at the near approximity of the electrode surface.

Although with both types of electrode arrangements (perpendicular or parallel to the flow) useful results have been reported, the wall-jet arrangment in some aspects seems to be superior.

C. Parameters Affecting the Transient Function of Ion-Selective Electrodes

The interpretation of dynamic response function of ISE in terms of experimental and membrane parameters, as well as the definition of response time, are key problems. With

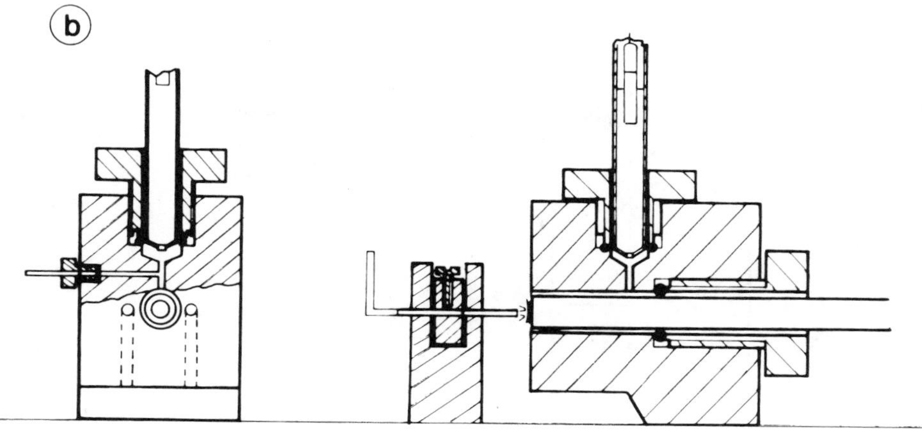

FIGURE 19b.

the aim of the above, the effect of the concentration of the primary ions; the presence of interfering ions and other nonionic solution constituents; the hydrodynamics prevailing in the potentiometric cell; and the ion-selective membrane composition and other membrane parameters were studied. The conclusions drawn from these studies were summarized in a review paper by Shatkay[134] and later by Lindner et al.[8,151]

1. The Primary Ion Concentration

When the activity step method is applied, the activity of the primary ion in the bathing electrolyte ($a_{i,1}$) is changed to a new stepped value ($a_{1,2}$):

$$a_{i,2} = b\ a_{i,1} \tag{41}$$

where $a_{i,1}$ and $a_{i,2}$ are the activities of the primary ion in the bulk of the sample solution prior to and following the activity step, while b is the proportionality factor.

The possible patterns of the activity change are shown in Figure 20.[152-153] In the first series of measurements (I), the ratio of the activities existing before and after the activity step was constant, while in the second case (II), the activity ratios varied, but either the initial ($a_{i,1}$) or the final activity value ($a_{i,2}$) was kept constant. In both types of studies, the proportionality factor b can either be smaller or larger than unity, depending on whether the effect of activity decrease or increase is aimed to study.

Concerning the effect of the solution activity on the transient functions or the response time (defined as a specially selected point of the latter; see Chapter 5), the following questions arise:

1. The effect of the absolute value of $a_{i,1}$ and $a_{i,2}$ on the transient signal when the ratio of the two activities is maintained constant. E.g., when this ratio is 10, the activity step can vary from 1×10^{-3} to 1×10^{-2} *M* or from 1×10^{-4} to 1×10^{-3} *M*. From electrode kinetics (see Section III), it follows that response time of an ISE should be concentration dependent[2] since all equations which describe the charging up process contain concentration data. In contrast to this, some authors report that changes in $a_{i,1}$ do not affect the response time in experiments, like in Figure 20 (series I[5,91,125,127,139,154]), while others note an increase in response time as $a_{i,1}$ is decreased.[4,24,124,134,136,140,152,153,155-162] Detailed studies on the response time of ISE in a large activity range (10^{-1} to 10^{-6} *M*) gave an answer to these contradictions (see Chapter 3, Section II).[8,152,153]

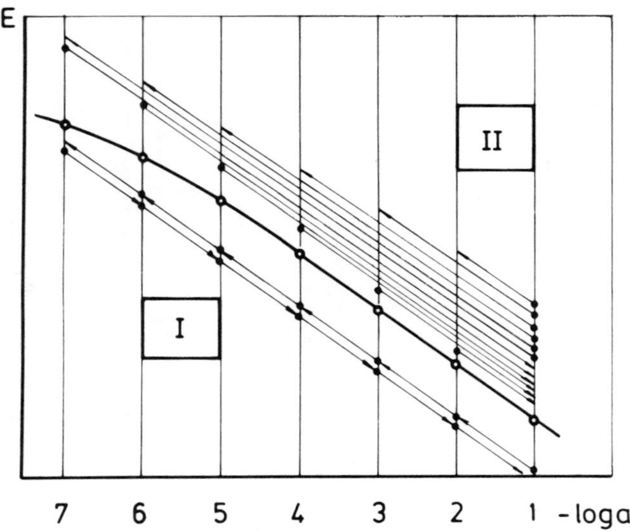

FIGURE 20. The activity steps used in the two series of measurements (I and II): (○) hypothetical calibration graph; (●) different $a_{i,1}$ values; (→) different $a_{i,2}$ values showing the direction of the activity change.

2. The effect of the change of the activity ratio, b, on the transient function. The ratio of the activities used in the experiments generally has a negligible effect,[91,125] especially when an activity increase is used:[163,164] e.g., when $a_{i,1} = 10^{-3}$ *M* and $a_{i,2}$ is either 10^{-2} *M* (b = 10) or 10^{-1} *M* (b = 100). However, at activity decrease, the t_{90} or t_{95} data are increasing with increasing activity ratios (e.g., when $a_{i,1} = 10^{-2}$ *M* and one applies an activity step to 10^{-4} *M* (b = 10^{-2}) or 10^{-5} *M* (b = 10^{-3}).[160,161,163,165]
3. The effect of the direction of the activity change. In general, from the studies it can be concluded that the response is always slower (flatter transient function) at an activity decrease compared to that observed at an activity increase.[4-8,91,134,145,152-154,160,161]

2. *Interferences*

In addition to primary ions, other ionic or neutral solution constituents can also affect transient function of ISE.

In a broad sense, one may consider as an interference the variation of response time due to conditioning,[123,124,137,166-170] leaching,[137] surface etching,[137] polishing,[6,171,172] the presence of interfering ions,[34,137,139,144,173-187] or of some nonionic electrode "poisons".[6,155,166,188,189] Besides these, the role of viscosity[160,161,169,170] and the ionic strength[5,169,170,190] of the analyte as well as of surfactants must also be considered.

The latter can effect the value and the transient function of other sources of potentials in the cell (e.g., diffusion potential, streaming potential), as well as the transport (diffusion) processes in the analyte through the change of viscosity and surface tension of the solution and the wetting of the electrode surface. Since the change in the cell voltage following a sudden change of the primary ion activity in the solution is the sum of all potential terms in the cell, thus all solution components other than the primary ion can affect the transient function measured in direct or indirect ways. However, due to the theoretical and experimental complexity of the problem, only a few publications can be found in the literature dealing with these questions.

3. *Hydrodynamic Conditions*

When applying the activity step method, the slowest step of the overall electrode process

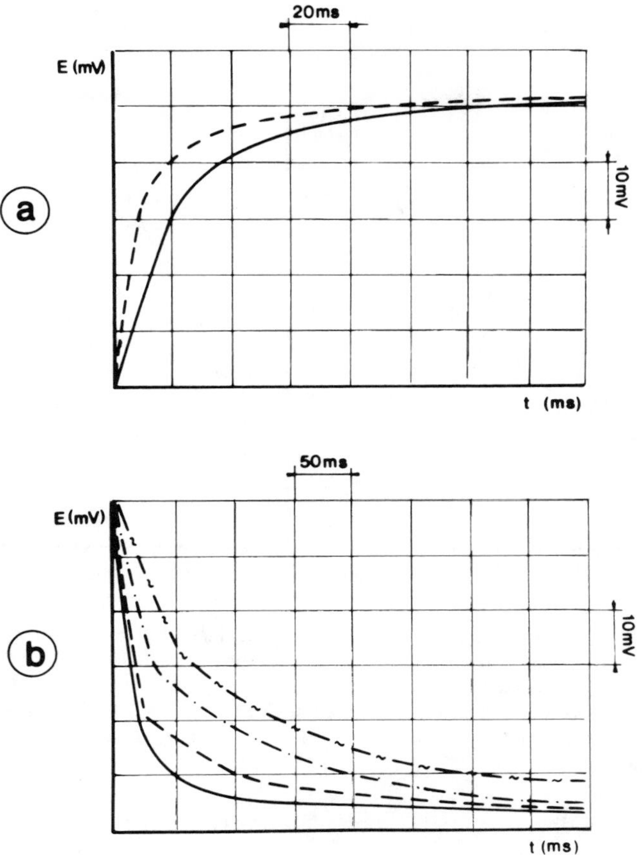

FIGURE 21. Effect of flow rate on the response time curves of an iodide-selective electrode (electrode membrane 1:1 Ag_2S + AgI ϕ = 12 mm). (a) $a_{i,1}$ = 10^{-3} *M* KI; $a_{i,2}$ = 10^{-2} *M* KI. (- - - -) v_1 = 117 mℓ/min; (——) v_2 = 54 mℓ/min. (b) $a_{i,1}$ = 10^{-2} *M* KI; $a_{i,2}$ = 10^{-3} *M* KI. (—∼—) v_1 = 30 mℓ/min; (–·–·–·) v_2 = 54 mℓ/min; (- - - -) v_3 = 117 mℓ/min; and (——) v_4 = 138 mℓ/min.

is usually the transport of the primary ion to the electrode membrane surface from the solution bulk; the transient function is controlled by diffusion processes in the adhering solution layer. This is reflected in the flow rate dependence of the transient function (Figure 21).[6-8,134,142,145,146,152,153,160,161] The distorting effect of the film diffusion which overlaps the transient function of the indicator electrode should be minimized. Accordingly, it is a requirement that the measuring cells used for response time studies[4-6,8,141,144,146,149] (Figures 15 to 19) ensure reproducible flow conditions and adhering layers as thin as possible.[6,8] The extrapolation of response time data recorded at a given flow rate to the "infinite" flow velocity (Figure 22)[145] does not necessarily give the "real" film diffusion-free response time of the electrode; however, it shows the capability of the measuring set-up.

When dynamic response studies are carried out in flowing or stirred solutions or using rotating or vibrating electrodes, the effect of the relative movement between the electrode surface and sample solution should also be considered on all potential terms arising in the cell.[134,191-194]

Van den Winkel and co-workers[194] studied the drift of baseline and the oscillation of potential ISE in flowing solutions. These phenomena were related to changes in streaming

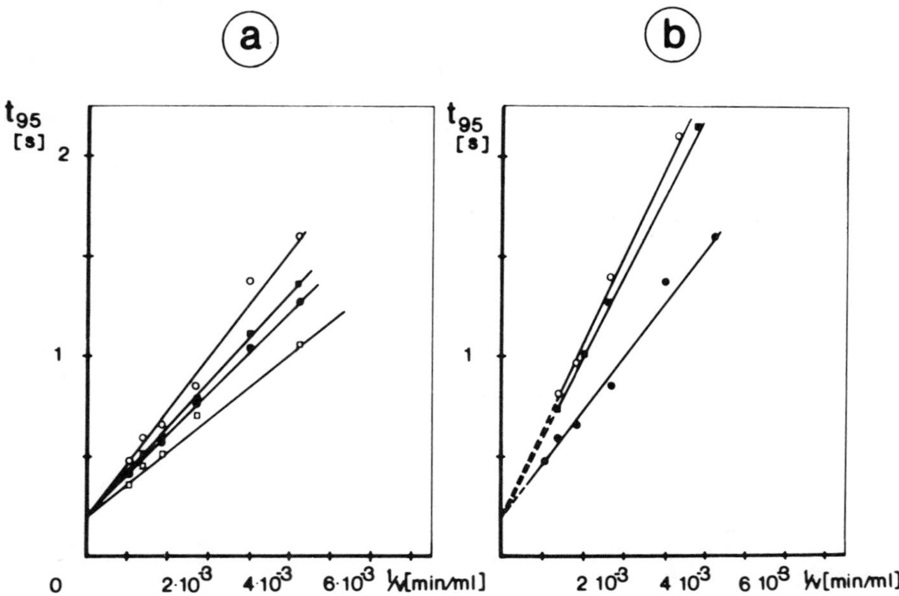

FIGURE 22. Plot of t_{95} vs. reciprocal flow rate. [t_{95} is the time required to attain 95% of the total potential change corresponding to the activity step (see Chapter 5, Section II)]. (a) At different concentration ratios and applying a Br^--selective electrode. (○) $c_1/c_2 = 10$; (■) $c_1/c_2 = 50$; (●) $c_1/c_2 = 100$; and (□) $c_1/c_2 = 1000$. (b) For three halide electrodes ($c_1/c_2 = 10$). (○) Cl^- electrode; (■) I^- electrode; and (●) Br^- electrode. (From Rangarajan, R. and Rechnitz, G. A., *Anal. Chem.*, 47, 324, 1975. With permission.)

potential in the small diameter tube section placed between the indicator electrode and the reference electrode. The effect of various experimental parameters, e.g., tube diameter, pumping frequency, and ionic strength, on the streaming potential and the methods of elimination of the latter were also reported. Unfortunately, no report was found on the transient functions of the streaming and diffusion potential and on the effect of flow rate on these functions.

4. *Membrane Bulk and Surface Parameters*

The chemical composition,[7,123,195,196] thickness,[4,115,155] membrane preparation technology,[195,196] and surface conditions[6,137,171,172,180] of the ISE have a pronounced effect on the transient functions. The alterations of the electrode surface during use[180] were also studied with different surface analytical techniques.[172,197-200]

The chemical and physical compositions of the membrane can influence the structure of the measuring surface,[201-204] the charge transfer reaction,[14] the recrystallization processes, the solubility of active material,[102,199,205] the surface tension between the contacting solution and the membrane, and, consequently, the thickness of the adherent solution layer. The membrane composition also affects the distribution processes between the membrane and the solution phases, etc.[96,97]

Each of the above processes may separately influence the dynamic properties of ISE, but often their overall effect shows up.

From all of the above, it follows that unequivocal conclusions can be reached only in some well-defined limiting cases due to the complexity of these processes (see Section III).

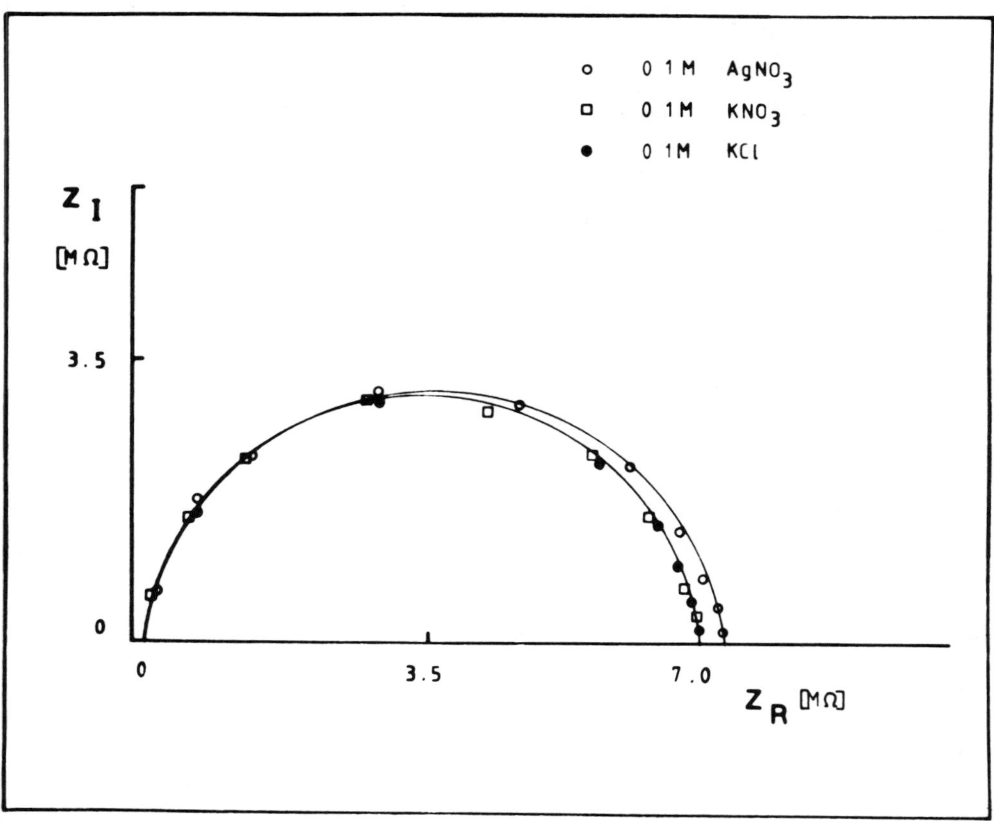

FIGURE 23. Complex impedance plane plots at 25°C of polished 0.64 mm. AgCl crystal in three bathing solutions: (○) 0.1 *M* $AgNO_3$; (□) 0.1 *M* KNO_3; and (●) 0.1 *M* KCl. (From Buck, R. P., *Hung. Sci. Instrum.*, 49, 7, 1980. With permission.)

VI. COMPARISON OF THE DIFFERENT TECHNIQUES

The various techniques dealt with in the previous sections are equally useful for the elucidation of the electrode response mechanism and the determination of electrode parameters of practical importance. The investigation of the electrode response mechanism may serve a practical purpose also, since the knowledge of the mechanism may permit the design of new types of ISE and the improvement of the already-known ones.

Polarization and impedance studies appear to be very promising as far as the comprehensive interpretation of electrode response mechanisms is concerned. In comparing the polarization and the impedance methods, the latter appears to have the advantage over the former, but neither technique permits the determination of the exchange current density of very rapid electrode processes at membranes of high resistance. Accordingly, only one single semicircle appears on the complex plane plots of AgCl membrane in either the presence or absence of the exchangeable ion in the bathing solutions (Figure 23).

The complexity of the impedance spectra of most widely used polycrystalline electrode membranes is another drawback of the technique. In such cases, the separation of the processes of time constants τ_B, τ_W, or τ_R (see Figure 5) can hardly be performed (Figure 24). The interpretation of the results is rendered especially difficult if the reproducibility of the measured data depends very much on the experimental conditions.

As the AC impedance method is regarded, it is rather discouraging that using small

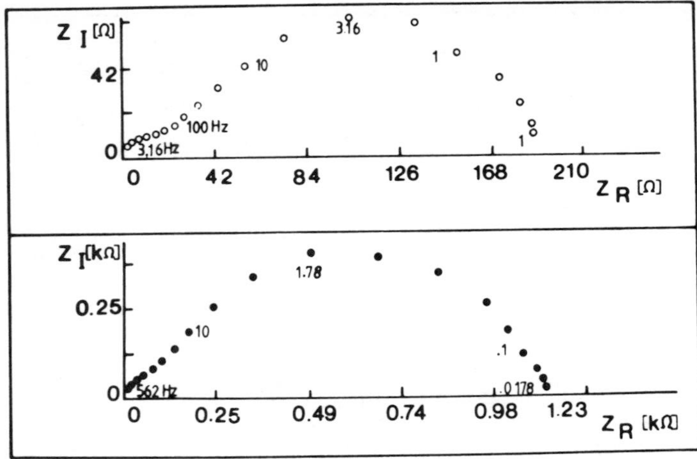

FIGURE 24. Impedance plane plots at a rotation rate of 36 rps in (○) 10^{-4} *M* $AgNO_3$ and (●) 10^{-5} *M* $AgNO_3$ for cold-pressed Ag_2S membrane (excess S^{2-}). (From Buck, R. P., *Hung. Sci. Instrum.*, 49, 7, 1980. With permission.)

amplitude perturbations — i.e., working in the linear part of current voltage curves — the differentiation between the exchange current density at a mixed potential point and at the real thermodynamic equilibrium is very difficult.[14] Furthermore, in the case of extremely low frequencies, the nearly DC current drawn through the membrane phase may change its chemical composition and, partly according to this and partly due to the asymmetry of current voltage curves, the frequency spectrum may be distorted, making any interpretation difficult.[14]

The activity step method is advantageous for the study of relatively slow processes. It is especially important from a practical point of view that the measurements are carried out at $j = 0$, i.e., under normal potentiometric conditions.

Moreover, in contrast to the techniques using a current intensity ($j \neq 0$), the activity step method and the corresponding transient function serve information about the processes taking place at one side of the membrane, i.e., membrane sample solution interface.[23]

With the polarization or the impedance methods the disturbing effects due to the polarization and the impedance of the reference and auxiliary electrodes cannot always be avoided — even by a judicious choice of the measuring setup and electrodes. The same considerations apply to the ohmic resistance of the solution, too.

Buck et al.[23] compared the impedance technique with the activity step method in the case of liquid membranes using digital simulation and obtained similar potential-time functions in both cases, though the two perturbation techniques are different. However, this comparison is straightforward only if the applied current density is smaller than the exchange current density.

In general, it has to be added that, from the transient signal recorded as an effect of an activity step realized under normal experimental conditions, usually fewer time constants can be determined than from the impedance diagrams. However, it is necessary to emphasize that with the suitable selection of experimental conditions, the activity step technique may give information on the processes of theoretical importance taking place at the solution-membrane interface.[152,153]

Besides, the transient functions determined by the activity step method are extremely important from practical points of view, as they give data to characterize the potentiometric cell which can be used to estimate the minimum time of analysis. From this aspect, the activity step technique cannot be substituted by any other method. Response time measurements are indispensable in order to characterize electrodes as well as newly designed analytical systems using potentiometric sensors as electrodes.

Chapter 3

THEORIES OF TRANSIENT POTENTIALS (FOLLOWING AN ACTIVITY STEP IN THE ABSENCE OF INTERFERING IONS)

I. INTRODUCTION

The voltage of a galvanic cell is equal to the sum of the individual potential terms arising in the cell assembly (Figure 1):

$$\underbrace{\underset{\epsilon_1}{\text{Ag, AgCl}}|\underset{\epsilon_2}{\text{dKCl}}\|\underset{\epsilon_3}{\text{salt bridge}}\|\underset{\epsilon_j}{\text{sample}}}_{\text{double-junction reference electrode}}$$

$$\underbrace{|\underbrace{\text{membrane}}_{\epsilon_M}|\text{internal solution}\underset{\epsilon_4}{|}\underset{\epsilon_5}{\text{AgCl, Ag}}}_{\text{ion-selective membrane electrode (indicator electrode)}} \quad (1)$$

$$\begin{aligned} E &= (\epsilon_1 + \epsilon_2 + \epsilon_3 + \epsilon_4 + \epsilon_5) + \epsilon_j + \epsilon_M \\ &= E^0 + \epsilon_j + \epsilon_M \end{aligned} \quad (2)$$

where E = cell voltage; E^0 = standard potential of the membrane electrode assembly, dependent on the composition of the internal solution and the reference electrodes, but independent of the sample solution composition; ϵ_j = potential contribution produced by diffusion layers within the aqueous system on the sample side (liquid junction potential); and ϵ_M = membrane potential.

In an ideal case, $\epsilon_j \approx 0$ and a Nernstian response of the membrane electrode cell is expected:

$$E = E_i^0 + \frac{RT}{z_iF} \ln a_i' = E_i^0 + S \log a_i' \quad (3)$$

where a'_i = activity of the ion I in the boundary layer of the sample solution in contact with the membrane surface; R = gas constant; T = absolute temperature; F = Faraday constant; z_i = charge of the primary ion; S = Nernstian slope; and $S = 2.303\ RT/z_iF = 59.16/z_i$ mV; 25°C.

It must be pointed out that the Nernst equation is clearly based on the assumption of thermodynamic equilibrium at the phase boundaries between membrane and outer solution. If thermodynamic equilibrium holds throughout the sample solution, then the usual relation applies:

$$E = E_i^0 + S \log a_i \quad (4)$$

where a_i = activity of ion I in the bulk of sample solution. Obviously, deviations from this form of the Nernst equation may arise due to time-dependent deviations between the surface activity a'_i and the intrinsic sample activity of the primary ion (a_i):

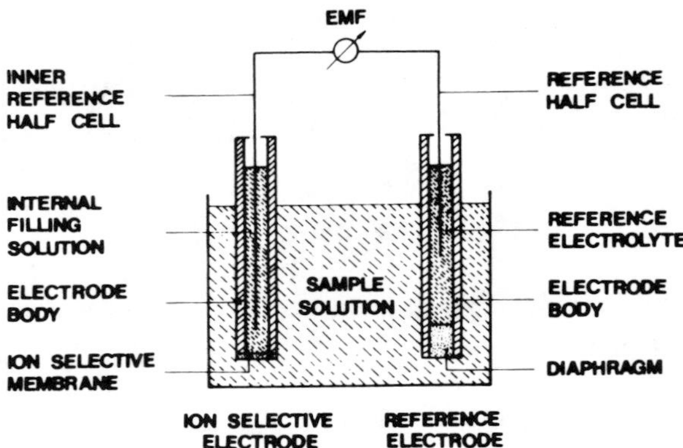

FIGURE 1. Schematic diagram of membrane electrode measuring circuit and cell assembly. (From Morf, W. E. and Simon, W., *Hung. Sci. Instrum.*, 41, 1, 1977. With permission.)

$$E(t) = E_i^0 + S \log a_i'(t) \tag{5}$$

where E(t) = the time-dependent cell voltage, and $a'_i(t)$ = the time-dependent activity of ion I in the boundary layer.

The time-dependent cell voltage following the activity step ($a_{i,1} \rightarrow a_{i,2}$) is, with a few exceptions, described with monotonous, asymptotic functions, starting from the initial cell voltage at t = 0, when

$$E_1 = E_i^0 + S \log a_{i,1} \tag{6}$$

until a new equilibrium is attained at $t \rightarrow \infty$:

$$E_2 = E_i^0 + S \log a_{i,2} \tag{7}$$

where E_1 is the equilibrium cell voltage in the bathing solution of activity $a_{i,1}$; E_2 is the equilibrium cell voltage following the activity step; and $a_{i,1}$ and $a_{i,2}$ are the activities of ion I in the bulk of the solution before ($t \leq 0$) and after ($t \geq 0$) the activity step, respectively.

The mathematical formulation describing the transient functions differs according to the assumed rate-determining partial process of the overall electrode reaction. The rate-determining step of the electrode process can be

1. The diffusion of the primary ion I to electrode surface from the bulk of the solutions[7,134,143]
2. Transfer of the potential determining components at the interface; this process can be affected by transport processes within the ion-selective membrane[7,65,66,140,161,206]
3. First- or second-order chemical reaction[5,24,25] generating ions which take part in the charge transfer reaction

It has already been mentioned in Chapter 2, Section V.A, that the transient functions of a galvanic cell incorporating the ion-selective indicator electrode can also be affected by the parameters of the measuring electronics (see Chapter 2, Figure 14).[128]

In the case of membranes of high resistance (e.g., microelectrodes[129-132], the transient function of the cell is determined primarily by the electrical measuring circuit.

Based on studies of simplified equivalent circuit models, the following equation was suggested[131,132] for calculating the potential response of a cell incorporating ISE after an activity step:

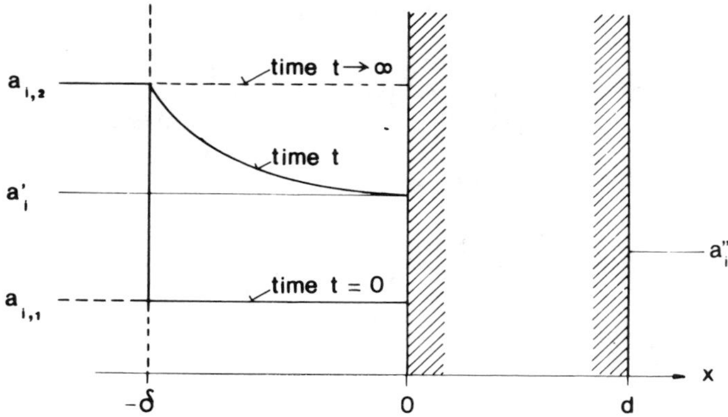

FIGURE 2. Diffusion model used for the ion-selective membranes of constant composition (glass, precipitate-based, and liquid ion-exchanger membranes). The course of the activity profile in the boundary layer after a step change in the sample activity at t = 0 is illustrated.

$$E(t) = E_2 - [E_2 - E_1]e^{-t/RC} \tag{8}$$

where R is the internal resistance of the galvanic cell, and C is the total capacitance of the cell and the amplifier input.

II. DIFFUSION THROUGH A STAGNANT LAYER

The considerable influence of the flow rate on the shape of the transient functions after an activity step (see Chapter 2, Figure 21) reflected the attention to the role of transport processes in the sample solution on the cell response.[6,7,134,142] In fact, the surface activity of the potential determining ion varies in function of time as a step change if the bulk ion activity is introduced.

Thus, diffusion processes will take place due to activity gradient at both sides of the membrane solution interface. Therefore, these processes at the end determine the transient function of a cell; at the same time, they may influence the lower limit of detection[92,207,208] and the selectivity factor measured.[180,183,184,209,210]

Steady-state diffusion through the stagnant solution layer was first described by Nernst[211] and later employed by Jaenicke[191-193] for describing the potential vs. time function of electrodes of second kind, and, more recently, Evans and others[212,213] applied it for ISE.[205]

A theoretical treatment of the dynamic characteristics, (i.e., the potential time function) of ISE incorporating solid-state, glass, or liquid ion-exchanger membranes (i.e., membranes of constant composition) was first presented by Markovic and Osburn[142] by considering steady-state diffusion through a stagnant solution layer. This model has been accomplished by Morf et al.,[7] and later by Shatkay,[134] for solving the time dependency of the measured surface activity, a'_i, illustrated in Figure 2.

It is apparent in Figure 2, that activity $a_{i,1}$ in the solution bulk and at the electrode surface a'_i are equal at $t \leq 0$. The activity of the sample solution is abruptly changed to $a_{i,2}$ at $t = 0$, at $x = -\delta$, outside the hydrodynamic boundary layer assumed to be of constant thickness. Thus, the change in time of the surface activity is determined by the diffusion through the hydrodynamic boundary layer.

An exact solution based on Fick's law can be found in the literature:[214,215]

$$a'_i = a_{i,1} + [a_{i,2} - a_{i,1}]\, f(t) \tag{9}$$

For relatively short times ($\sqrt{D't} \ll \delta$), f(t) can be given as follows:

$$f(t) = 2\left\{\sum_{n=0}^{\infty} (-1)^n \operatorname{erfc} \frac{(2n+1)\,\delta}{2\sqrt{D't}}\right\} \tag{10a}$$

while, for relatively long times ($\sqrt{D't} \gg \delta$):

$$f(t) = 1 - \frac{4}{\pi}\sum_{n=0}^{\infty} \frac{(-1)^n}{2n+1} \exp\left[-\frac{D'\pi^2}{4\delta^2}(2n+1)^2 t\right] \tag{10b}$$

where D′ is the diffusion coefficient; δ is the thickness of the stagnant solution layer; and

$$\operatorname{erfc}(z) = 1 - \operatorname{erf}(z) \tag{11}$$

$$\operatorname{erfc}(z) = \frac{2}{\sqrt{\pi}}\int_o^z e^{-\eta^2}\, d\eta \tag{12}$$

$$\eta = \xi/2\sqrt{D't} \tag{13}$$

where ξ is a distance from a given point.[214,215] The erf (error function) has the properties:

$$\operatorname{erf}(-z) = -\operatorname{erf}(z) \qquad \operatorname{erf}(0) = 0 \qquad \operatorname{erf}(\infty) = 1 \tag{14}$$

The difference between the short- and long-time behavior of the transient function of ISE is expressed mathematically by Equations 10a and 10b. The same conclusion was drawn by Buck[9,22] from the origin of finite Warburg behavior of ion-selective membranes in which the potential can be approximated, with a $t^{1/2}$ function at short times, but with an exponential one at long times.

It is sufficient for practical purposes to use the first term in the summation in the time range $\sqrt{D't} \ll \delta$ (Equation 10a) or $\sqrt{D't} \gg \delta$ (Equation 10b). The extent of simplification was controlled by model calculations (Figures 3 and 4), thus Equations 10a and 10b were simplified:

$$a_i' = a_{i,1} + 2[a_{i,2} - a_{i,1}] \operatorname{erfc} \frac{\delta}{2\sqrt{D't}} \tag{15}$$

and

$$a_i' = a_{i,1} + [a_{i,2} - a_{i,1}]\left[1 - \frac{4}{\pi}\exp\left(-\frac{D'\pi^2}{4\delta^2}\right)t\right] \tag{16}$$

or

$$a_i' = a_{i,1} + [a_{i,2} - a_{i,1}]\left[1 - \frac{4}{\pi}\exp\left(-t/\tau'\right)\right] \tag{17}$$

where the τ′ time constant:

$$\tau' = \frac{4\delta^2}{\pi^2 D'} \tag{18}$$

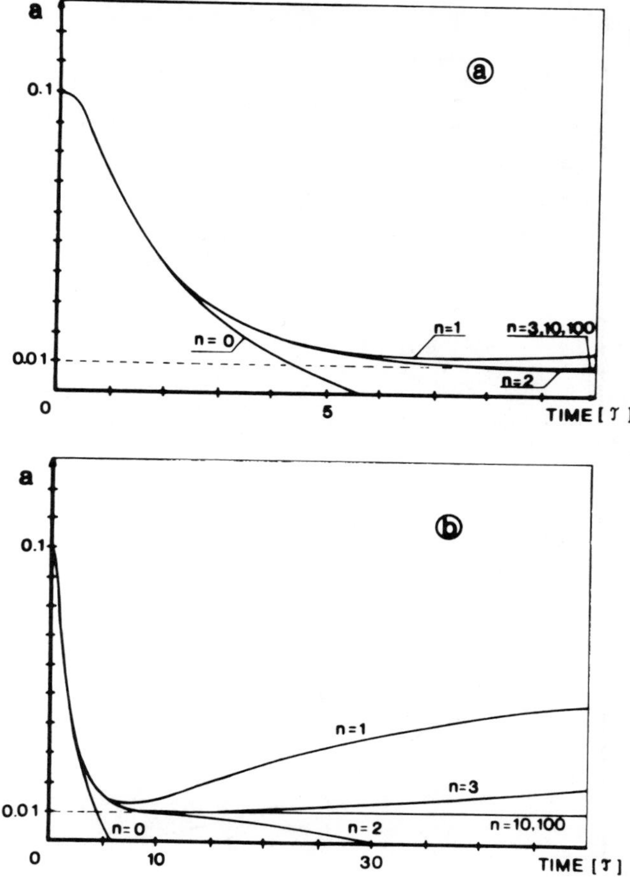

FIGURE 3. Graphical presentation of the surface activity change in time following a one-decade activity decrease. The curves were calculated with Equations 9 and 10a and plotted at different numbers of terms (n) in two time domains ($a_{i,1} = 10^{-1}$ *M*; $a_{i,2} = 10^{-2}$ *M*): (a) time range 0 to 0.5 τ; (b) time range 0 to 3.0 τ.

For a semiquantitative treatment in the entire time range, the following simplified expression was used:

$$a_i' = a_{i,1} + [a_{i,2} - a_{i,1}]\,[1 - e^{-t/\tau'}] \tag{17a}$$

with the time constant:

$$\tau' \approx \frac{\delta^2}{2D'} \tag{18a}$$

By inserting Equation 17 into Equation 5, the following approximation for the time-dependent cell potential is obtained:

$$E(t) = E_i^o + S \log \{a_{i,2} - [a_{i,2} - a_{i,1}]e^{-t/\tau' + 0.24}\}$$

$$= E_2 + S \log \left[1 - \frac{a_{i,1}}{a_{i,2}}\, e^{-t/\tau' + 0.24}\right]$$

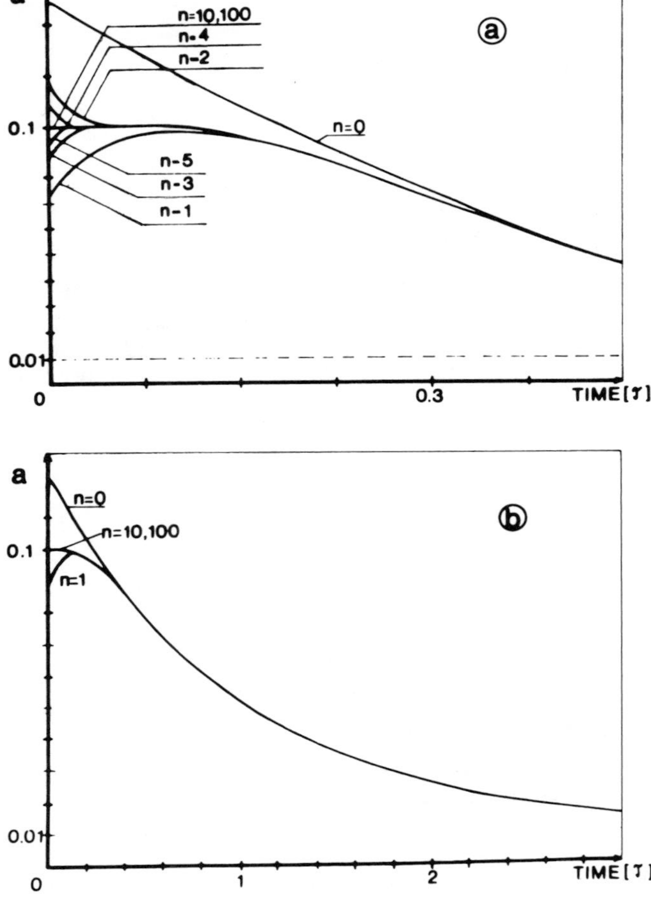

FIGURE 4. Graphical presentation of the surface activity change in time following a one-decade activity decrease. The curves are calculated with Equations 9 and 10b and plotted at different numbers of terms (n) in two time domains ($a_{i,1} = 10^{-1}$ *M*; $a_{i,2} = 10^{-2}$ *M*): (a) time range 0 to 10 τ; (b) time range 0 to 50 τ.

$$= E_I + S \log \left[1 + \frac{a_{i,2} - a_{i,1}}{a_{i,1}} \left(1 - e^{-t/\tau' + 0.24} \right) \right] \tag{19}$$

If the activity step is sufficiently small, the resulting voltage is less than RT/zF V (i.e., the perturbation is in the so-called linear range):

$$\frac{a_{i,2} - a_{i,1}}{a_{i,1}} \left[1 - \frac{4}{\pi} \exp\left(-\frac{t}{\tau'} \right) \right] << 1 \tag{20}$$

the logarithmic expression in Equation 19 can be expanded in series and the higher-order terms can be neglected; thus, Equation 19 is rewritten in simple form:

$$E(t) = E_I + S \frac{a_{i,2} - a_{ai,1}}{a_{i,1}} (1 - e^{-t/\tau'}) \tag{21}$$

The limiting values of term $[1 - \exp -t/\tau']$ are zero at $t = 0$ and it is unity at $t = \infty$. Thus, the term $[(a_{i,2} - a_{i,1})/a_{i,1}]$ determines whether the approximation is justified.[17] (See also Chapter 2, Section II.C.) For small activity steps, Equation 21 should fit along the transient, while the approximation will hold only during the initial stages of the transient for large values of $[(a_{i,2} - a_{i,1})/a_{i,1}]$ (outside the linear regime).

Equivalent expressions with Equations 19 and 21, but with different time constants[46] or without the factor $[1 - a_{i,1}/a_{i,2}]$,[5,140,142,206] were derived earlier. It was shown that these and similar exponential relationships (see Equation 54) give a close fit of the dynamic response curves observed for membranes of constant composition[5,153,206] (Figure 5) as long as the effect of interfering ions, and the diffusion of the ionic species within the membrane bulk, can be neglected.

Since ionic species are in an approximately constant concentration within the membrane bulk (given by the concentration of ion-exchange sites), thus ionic diffusion through the membrane becomes negligible under potentiometric conditions.[7,17,86,87,132] Therefore, it is the diffusion within the boundary layer of the sample solution which mainly determines the speed and mode of electrode response.[7,134,142,145]

Finally, the response time of the ion-selective sensors can be expressed as a function of the residual relative deviation in the cell voltage:[7]

$$t(\epsilon) = \tau'\left[\ln \frac{1 - \dfrac{a_{i,1}}{a_{i,2}}}{1 - \left[\dfrac{a_{i,1}}{a_{i,2}}\right]^{\epsilon}} + 0.24\right] \tag{22}$$

where ϵ is the residual relative deviation in the cell voltage (E):

$$\epsilon = \frac{E(t) - E_2}{E_1 - E_2} \tag{23}$$

A comparison of numerical response time values corresponding to different definitions and experimental conditions is given in Table 1 (see also Chapter 5).

It is apparent from Equations 18, 19, and 22 that the thickness of the diffusion layer (δ) is the most important parameter, that determines the dynamic response characteristics of ISE based on membranes of constant composition. The thickness of the aqueous diffusion layer depends on the shape and condition of the electrode surface (surface morphology[6,8]) and on the composition of the sample solution (viscosity, surface wetting, etc.).[160,161,169,170,193] The dimensions of the stagnant solution layer can be reduced drastically by stirring or removing pores and impurities from the membrane surface.[6,8,160,161] With glass electrodes, reduction of the hydrated surface layer by etching can sometimes prevent sluggish response behavior.[137,138]

The thickness of the hydrodynamic boundary layer can be increased considerably if the electrode is used in strongly corrosive media (Figure 6) causing deep scratches on the electrode surface.[6,172,197-199,205]

The dynamic response characteristics of ISE depend to a great extent on the direction of the activity change in the solution. It is clearly evident from Figure 7 and Table 1 that the response time increases considerably at an activity decrease compared to that of an activity increase. Moreover, it follows from Equations 19 and 22 that this effect should be independent of the primary ion concentration or, in other words, of the "activity level". (See also Chapter 2, Figure 20 (series I); Section II.A.)

By considering Equations 22 and 23, the report of Rangarajan and Rechnitz[145] (Chapter 2, Figure 22) is justified in that t_{95} increases parallel with the decrease of the activity ratio.

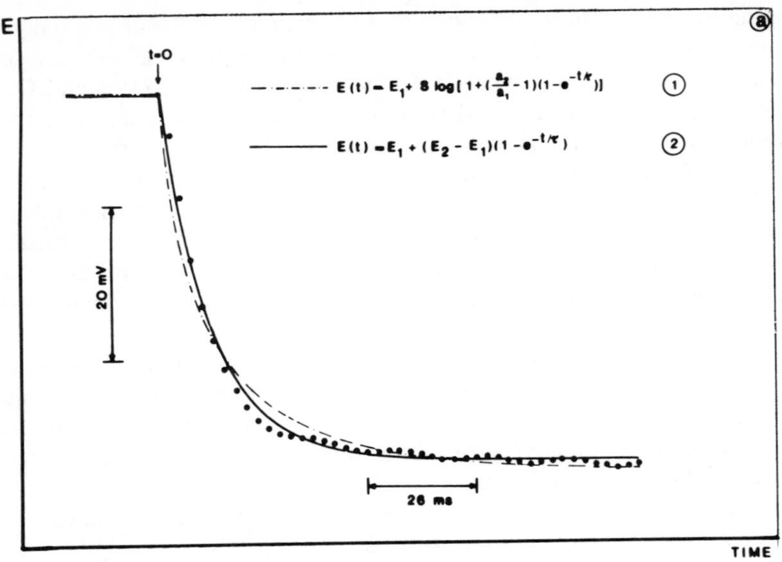

FIGURE 5. Comparison of the dynamic response curves recorded experimentally and calculated by Equation 19 (1) or Equation 54. (2) in two different time domains. Electrode: Radelkis OP-7112-I, AgI based iodide selective electrode; Activity step: 10^{-3} *M* KI → 10^{-4} *M* KI; Ionic strength: 0.1 *M* KNO_3 Measuring set up: Chapter 2, Figure 19. Fitted parameters:

Equation 19	Equation 54
Figure 5a	
S = −48.24 mV/decade	$E_2 - E_1$ = −46.79
τ = 26.91	τ = 12.17
RSS = 170	RSS = 43
RMSD = 2.35 mV	RMSD = 1.0 mV
Figure 5b	
S = −48.28	$E_2 - E_1$ = −48.11
τ = 25.64	τ = 12.88
RSS = 28	RSS = 29.64
RMSD = 0.68	RMSD = 0.74

$$\text{where RSS} = \sum_{i=1}^{N}(E_{meas} - E_{calc})^2$$

and

$$\text{RMSD} = \sqrt{\frac{\text{RSS}}{N - p}}$$

and p is the number of parameters adjusted in the minimization procedure.

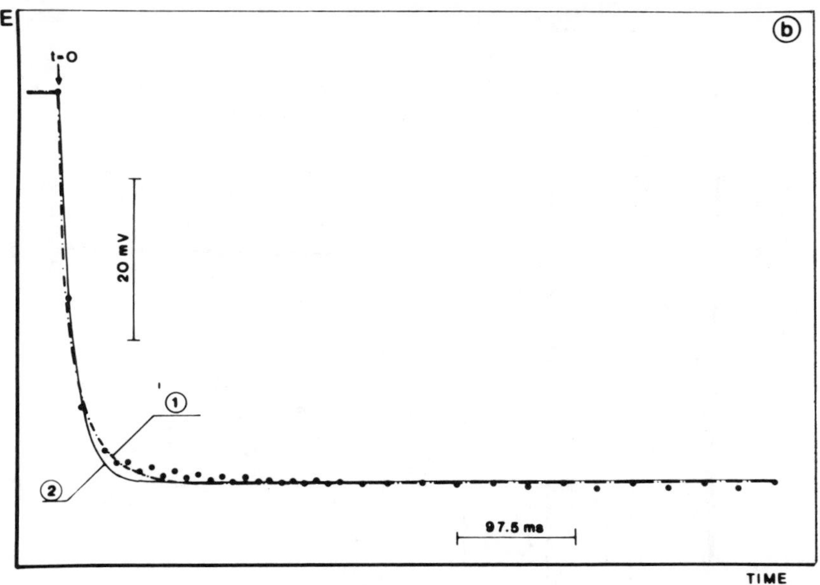

FIGURE 5b.

Table 1
THEORETICAL RESPONSE TIME VALUES FOR MEMBRANES OF CONSTANT COMPOSITION CALCULATED ACCORDING TO EQUATION 19

Response time parameter		Values (sec) calculated with $\tau' = 1$ sec
$t_{1/2}$	For tenfold	0.52
t_{95}	activity	2.35
$t_{99.5}$	increase	4.61
$t_{1/2}$	For tenfold	1.67
t_{95}	activity	4.54
$t_{99.5}$	decrease	6.90

Note: $t_{1/2}$, t_{95}, $t_{99.5}$ are times required to attain 50, 95, 99.5% of the total potential change relevant of the activity step (see Chapter 5, Section II).

A complication arises if an additional potential difference is produced within the boundary layer by the differing diffusion rate of ions in the sample. This is generally the case for sample cations I and anions X having different diffusion coefficients, D'_I and D'_x.[7] The flux of both ionic species may be represented by Fick's law only if a mean diffusion coefficient is defined:

$$D' = \frac{(z_I - z_x)\, D'_I D'_x}{z_I D'_I - z_x D'_x} \tag{24}$$

By including the diffusion potential term:

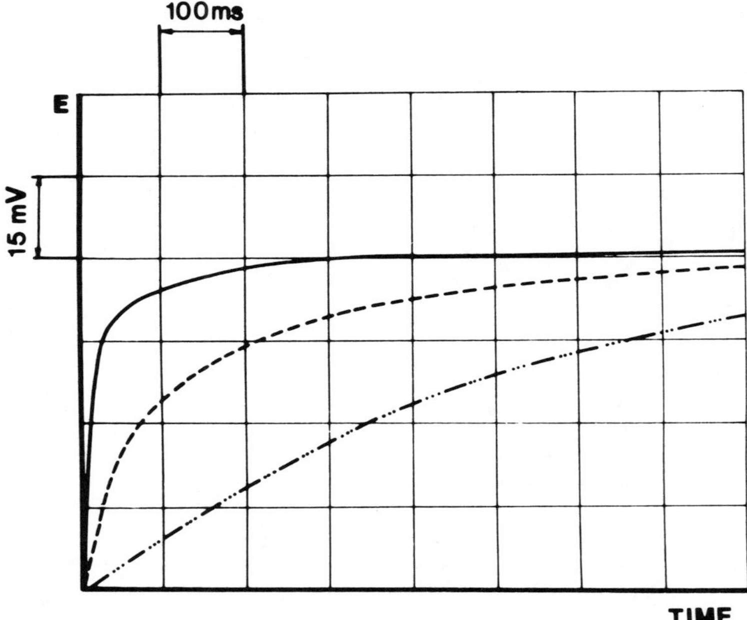

FIGURE 6. Response time curves of a silicone rubber iodide-selective electrode (based on AgI) as the function of soaking time in 10^{-1} *M* CN^- solutions: $a_{i,1} = 10^{-3}$ *M* KI, $a_{i,2} = 10^{-2}$ *M* KI, v = 54 mℓ/min, ø = 6 mm. (——) Without pretreatment; (– – – – –) after soaking for 5 min; and (—···—) after soaking for 10 min.

$$\Delta E' = \frac{D'_I - D'_x}{z_I D'_I - z_x D'_x} \cdot \frac{RT}{F} \ln \frac{a_{i,2}}{a'_i}$$

$$= \left(\frac{D'}{D'_I} - 1\right) S \log \frac{a'_i}{a_{i,2}} \qquad (25)$$

the dynamic response is obtained by the following general equation:

$$E(t) - E_2 = \frac{D'}{D'_I} S \log \frac{a'_i}{a_{i,2}} \qquad (26)$$

Obviously, the simplified relationship given in Equation 19 is only valid for binary electrolytes with $D' \cong D'_I$ or for sample solutions with an inert electrolyte added in order to eliminate the diffusion potential within the boundary layer (e.g., KCl or KNO_3). Thus, it is not surprising that noninterfering ions may also have a certain effect on the potential time profile of a sensor.[5,169,170,190] Probably, the nonmonotonic transient response of noninterfering cations observed in the two-ionic range of glass electrodes (see Chapter 4) may also be explained partly in terms of a potential difference across the boundary layer (hydrated layer on the glass surface). In fact, potentials of the type shown by Equation 25 are clearly transient ones.

The response time in general (Equation 22) is drastically increased by using sample solutions which induce extensive diffusion processes within the membrane surface. Long delays in the electrode response have been observed in two cases: (1) in the presence of interfering ions, which has an effect on the membrane surface com-

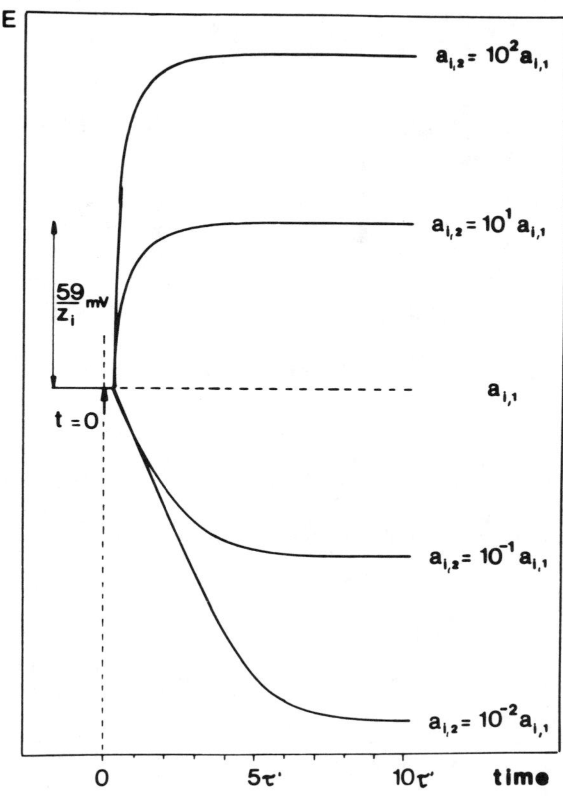

FIGURE 7. Potential response vs. time profiles for ion-exchange membrane electrodes, calculated according to Equation 19.

position,[137-139,180-187,191-193,209,210] and (2) in the region of the lower detection limit[25,134,136,152,153,156-159] where dissolution reactions occur (see Section II.A).[25,100,101]

An equation of the same type as Equation 19 can also be adapted as a first approximation for the transient response of membrane-covered electrodes such as gas-sensitive electrodes.[6,7,216-218] This area has been studied in detail by Morf et al.[219] and others.[220,221] The gas-permeable membrane represents an external diffusion layer. Similarly, the sluggish response of some silicone rubber-based electrodes can be attributed to a high resistance silicone rubber layer covering the electrode surface (Figure 8) as a consequence of membrane preparation technology.[6,222] At the latter, the high resistance film serves as an additional diffusion layer. Accordingly, the effective thickness of the diffusion layer increases, and the mean diffusion coefficient decreases remarkably. This assumption is supported by the fact that, in contrast to other types of electrodes studied under the same experimental conditions, the rate of response of silicone rubber-based potassium electrodes is practically unaffected by the flow rate of the sample solution (Figure 9), i.e., by the thickness of the stagnant diffusion layer in the solution.[222] (cf. Chapter 2, Figure 21, and Figure 9.) Moreover, the time constants (Equation 18) obtained by fitting Equation 19 to the experimental data have been found unexpectedly high, which means that, according to Equation 18, the effective diffusion layer thickness is much larger than that of the unstirred aqueous boundary layer (Figure 10). In the case of precipitate-based ISE, this thin, high resistance silicone rubber layer is usually polished off to guarantee a fast electrode response.[6,197]

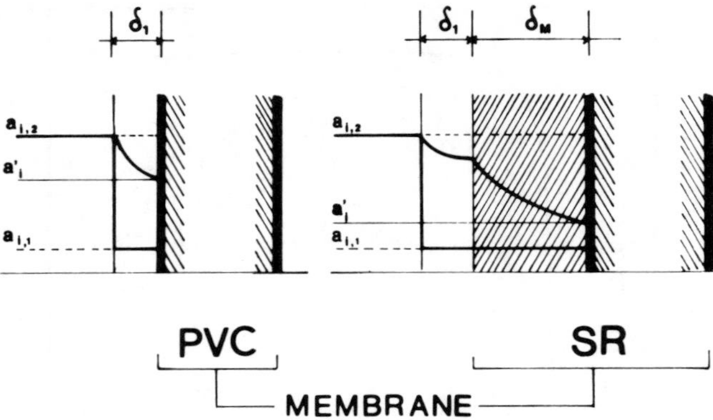

FIGURE 8. A simplified model to show the concentration profile at the membrane surface of a silicone rubber-based electrode following an activity step. δ_l = the thickness of the unstirred boundary layer (diffusion layer thickness). δ_M = the thickness of adhering silicone layer covering the electrode surface (thus, the diffusion layer thickness is the sum of δ_l and δ_M).

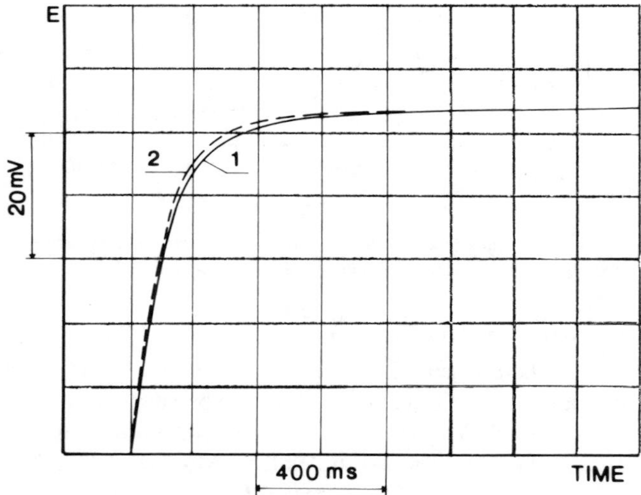

FIGURE 9. Effect of flow rate on the response time curve of a silicone rubber-based valinomycin potassium ISE: $a_{i,1} = 10^{-4}$ *M* KCl, $a_{i,2} = 10^{-3}$ *M* KCl. (1) (—) v = 55 mℓ/min; (2) (----) v = 135 mℓ/min.

A. Limits of the Diffusion Model

The model based on the assumption that the diffusion through the hydrodynamic boundary layer is the rate-determining process of the electrode response led to the somewhat negative conclusion that, at rapid interfacial reactions, the measurement of transient functions with the activity step method only provides information on diffusion phenomena in the hydrodynamic boundary layer, or on the performance of the measuring instrument, and not on the ion-selective membrane properties.

In order to study the electrode process on the ion-selective membrane following the diffusion from the solution to the interface, first Pungor et al.[6,152,153] investigated the deviations from the diffusion model systematically.

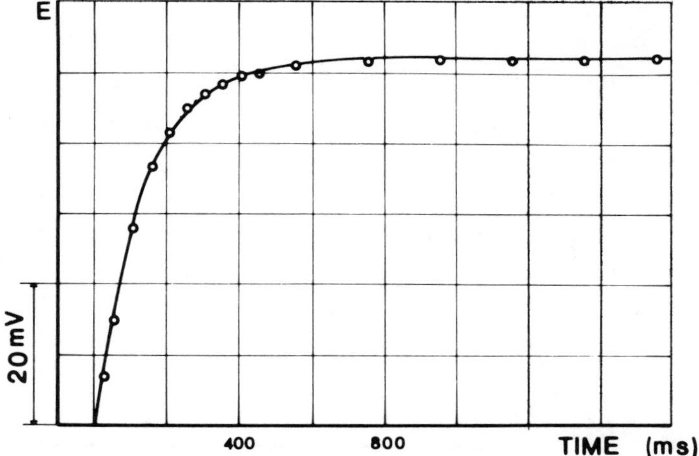

FIGURE 10. Calculated and measured response time curves of a silicone rubber-based potassium ISE: $a_{i,1} = 10^{-3}$ *M* KCl, $a_{i,2} = 10^{-2}$ *M* KCl. (——) Recorded curve; (○○○) values calculated with Equation 19 ($\tau' = 225$ msec).

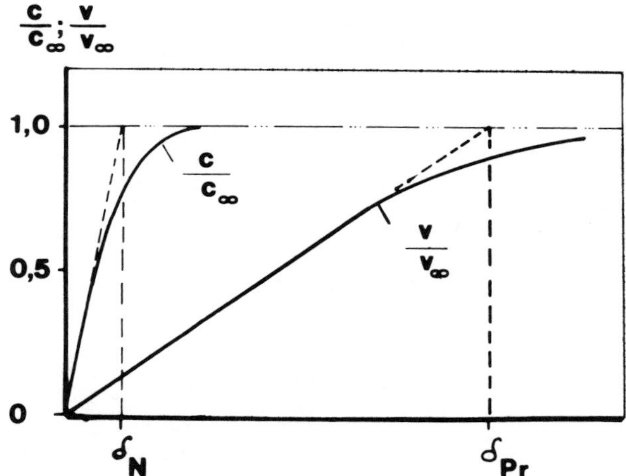

FIGURE 11. The concentration and the laminar flow rate profiles in the hydrodynamic boundary layer. (From Vielstich, W., *Z. Electrochem.*, 57, 646, 1953. With permission.)

By the appropriate selection of experimental conditions, it can be achieved that the transient functions are not distorted by diffusion processes through the stagnant boundary layer and, thus, the rate of the electrode response is determined by interfacial processes characteristic of the ion-selective membrane. For this, the thickness of the Nernst diffusion layer (δ_N) and that of the hydrodynamic layer (Prandtl layer) (Figures 11 and 12) had to be calculated as a first step.[223]

The relationships derived for the calculation of the thickness of the hydrodynamic (δ_{Pr}) and the Nernstian (δ_N) boundary layers depend first of all on the types of flow (laminar or turbulent) in the fluid as well as on the geometry of the electrochemical cell.

The following expressions are valid for laminar flow parallel to the electrode surface:[2,223,224]

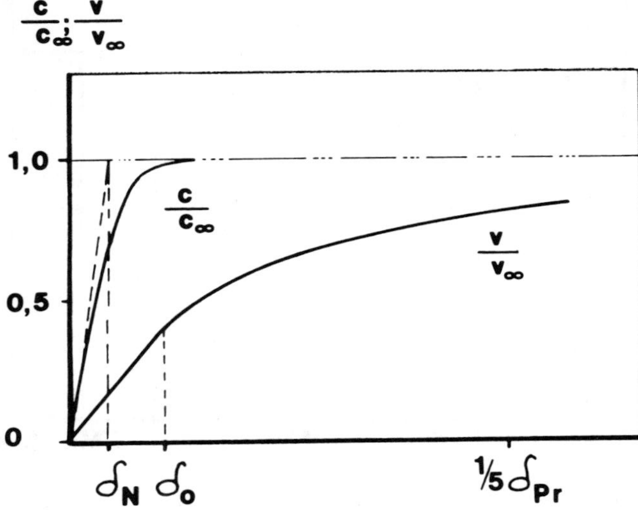

FIGURE 12. The concentration and the turbulent flow rate profiles in the hydrodynamic boundary layer. (From Vielstich, W., *Z. Electrochem.*, 57, 646, 1953. With permission.)

$$\delta_N = \delta_{Pr} \cdot Pr^{-1/3} \tag{27}$$

and

$$\delta_N = 31^{1/2} v^{-1/2} \nu^{1/6} D'^{1/2} \tag{28}$$

while in the case of turbulent flow:

$$\delta_N \sim 1\ Re^{-0.9} Pr^{-1/3} = \delta_0 Pr^{-1/3} \tag{29}$$

$$\delta_N = 1^{0.1} v_\infty^{-0.9} \nu^{17/30} D'^{1/3} \tag{30}$$

where V_∞ is the flow velocity of the solution flowing parallel to the electrode surface in the direction of coordinate 1, at infinite distance from the surface (m/sec); ν is the kinematic viscosity (m^2/sec); D' is the diffusion coefficient in the adhering layer (m^2/sec);1 is the coordinate in the direction of flow (m) (see Chapter 2, Figure 16); Pr is the Prandtl number (called also Sc: Schmidt number):

$$Pr = \frac{\nu}{D'} \tag{31}$$

Re is the Reynolds number:

$$Re = \frac{v \cdot 1}{\nu} \tag{32}$$

and δ_o is the linear range of the velocity profile or the thickness of the viscous boundary layer (cf. Figure 12).

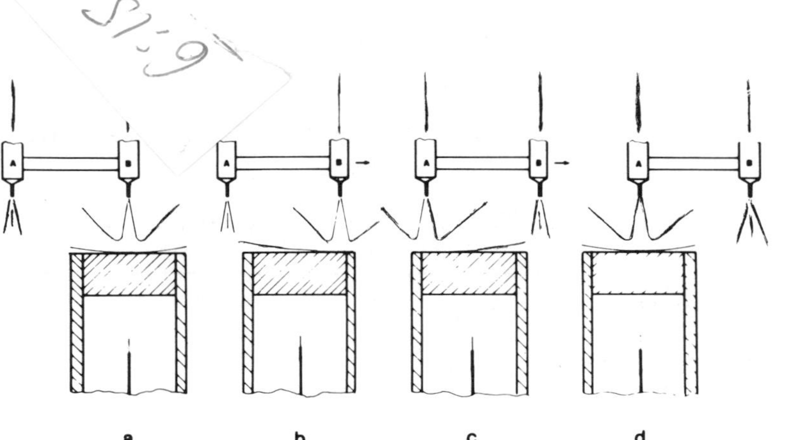

FIGURE 13. Flow profile variation at the electrode surface as the jet position is changed (see also Chapter 2, Figure 16).

The expression derived for the Nernstian diffusion layer (δ_N) in a wall-jet cell arrangement used in our experiment (Chapter 2, Figures 16 and 19) differs from Equation 30 by a multiplying factor, and also by the value of the Reynolds number due to the different cell geometry:[225]

$$Re = \frac{d_n v_{STR}}{\nu} \tag{33}$$

where v_{STR} is the flow rate parallel to the surface in wall-jet cell arrangement:

$$v_{STR} = \frac{6.4 \cdot v_x \cdot d_n}{x + 1} \tag{34}$$

where v_x is the linear flow rate perpendicular to the electrode surface; x is the distance between the jet orifice and the electrode; and d_n is the diameter of the jet.

It is apparent from Equations 27 and 29 that, in aqueous solutions ($D' \sim 1 \times 10^{-9}$ m^2/sec; $\nu \sim 10^{-6}$ m^2/sec), the thickness of the hydrodynamic boundary layer and that of the viscous boundary layer in the case of turbulent flow are about ten times larger than the thickness of the Nernstian diffusion layer:[223]

$$\delta_{Pr} \sim 10\delta_N \tag{35}$$

and

$$\delta_0 \sim 10\delta_N \tag{36}$$

In practice, steady-state flow conditions are generally assumed, i.e., the flow velocity at various points of the flow profile is supposed to be constant and, consequently, steady-state concentration profiles are formed also in the Nernstian layer.

The potential vs. time relationship given by Equation 19 was derived on the basis of the above assumptions. It has to be noted, however, that during the activity step: (1) flow conditions change continuously (Figure 13);[151](2) the flow velocity can vary; and (3) the viscosity of the solution may vary on account of the change in concentration. Following the activity step, the new hydrodynamic boundary layer develops, generally in a very short time,[220] and consequently the Nernstian diffusion layer changes also. The thicknesses of

Table 2
CORRECTED INITIAL SLOPE (m^*_{eff})[a] OF RESPONSE TIME CURVES

	Recorded in Ag^+ solutions (mV/msec)		Recorded in I^- solutions (mV/msec)	
Electrode type	**25°C**	**65°C**	**25°C**	**65°C**
Ag metal	5.5	7.5		
Ag_2S	4.8			
AgI + Ag_2S	5.9		4.1	7.0
AgI	3.7	8.3	3.7	8.1
AgI (SR)[b]	5.5	9.4	3.3	7.8
$\overline{m}_{eff}$[c]	5.1	8.4	3.7	7.6

[a] $m^*_{eff} = m \dfrac{59.16}{E}$, where m is the experimentally determined slope of the response time curves, and E is the potential coresponding to the activity step.
[b] SR = silicone rubber matrix.
[c] $\overline{m}_{eff}$ = the average m_{eff} value.

both the hydrodynamic (δ_{Pr}) and the Nernstian (δ_N) boundary layers were estimated from the geometrical parameters of the measuring set-up (Chapter 2, Figure 16), taking into account the experimental conditions.

For a laminar flow, $\delta_N = 6.5 \times 10^{-7}$ m (Equation 28), while for a turbulent flow, $\delta_N = 1. \times 10^{-7}$ m (Equation 30) were obtained. From the latter, it follows that the thickness of the viscous boundary layer, δ_o, is 1×10^{-6} m. The conditions used for the calculation were $l = 7 \times 10^{-3}$ m, $v = 3$ m/sec, $\nu = 10^{-6}$ m²/sec, $D' = 2 \times 10^{-9}$ m²/sec, which were characteristic for the experiments of Lindner et al.[6,8]

The calculated value of the boundary layer thickness was found to be thinner than the surface roughness of membranes prepared by pressing analytical precipitates (grain size of 2 to 3 μm).[161] Similary, the value of τ' (Equation 18a) calculated from the layer thickness (δ_o):

$$\tau' \sim \frac{(10^{-6})^2}{2.10^{-9}} = 5.10^{-4} \text{ sec}$$

is some orders of magnitude smaller than the data obtained by curve fitting (Figure 5).[226] These may suggest that under experimental conditions, i.e., $v \geqslant 3$ m/sec, $l \leqslant 7 \times 10^{-3}$ m, the shape of the dynamic response curve and thus the value of response time depend not only on diffusion processes, but other reactions contributing to the overall electrode response may also show up on the transient signals.

This assumption was also supported by the fact that the response of a silver iodide-based electrode was found to be more rapid in silver nitrate solutions than in potassium iodide solutions using equal activity steps in both cases (Table 2).[6,8,160,161] This phenomenon cannot be explained by a diffusion model because the opposite results would be expected according to the diffusion model by considering the diffusion coefficients ($D'_{AgNO_3} = 1.768 \cdot 10^{-9}$ m²/sec; $D'_{KI} = 2.001 \cdot 10^{-9}$ m²/sec)[227] of the respective salts or the relative ion mobilities of silver and iodide ions in the adhering layer ($U_{Ag^+} = 50.2$ cm²/Ω; $U_{I^-} = 62.7$ cm²/Ω).[22]

Denks and Neéb[160,161,226] also rejected the validity of the film diffusion model at very high flow velocities applied in their experiments ($v = 6.8$ m/sec; $\delta_N^{Turb} \sim 6 \times 10^{-8}$ m; $\delta_o \sim 6 \times 10^{-7}$ m; $\tau \sim 5 \times 10^{-6}$ sec).

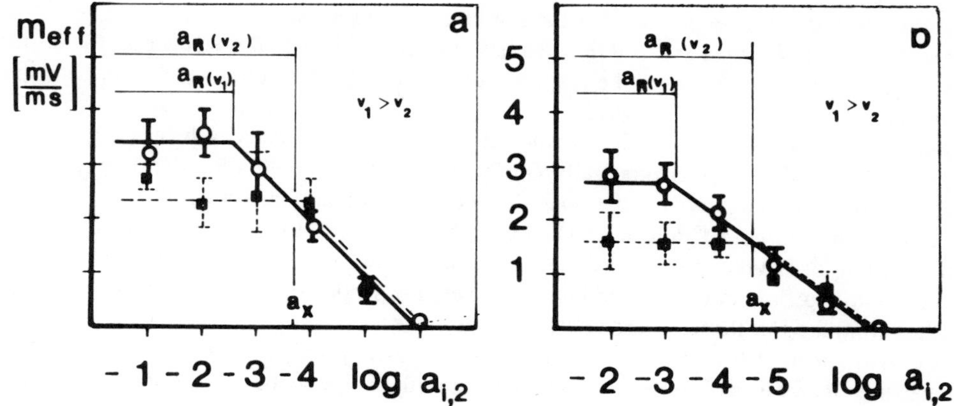

FIGURE 14. The normalized initial slope values as a function of log $a_{i,2}$ in the first series of measurements (I in Chapter 2, Figure 20): (a) in the direction of concentration increase; (b) in the direction of dilution; (○) flow rate v_1 = 140 mℓ/min; (⊠) flow rate v_2 = 100 mℓ/min. Each of the data measured are given along with their standard deviation: a_R = the activity value independent ($m_{eff} \neq f(a_{i,2})$) range; a_x = the critical activity value below which the normalized slopes are flow rate independent.

Thus, on the basis of the foregoing considerations, deviations from the diffusion model can be interpreted.

1. The Effect of the Activity Level

Obviously, the electrode processes following the diffusion can be studied only by designing experimental conditions under which the rate of the ion transport from the solution to the electrode surface is higher than that of the consecutive reactions. Such conditions are met, e.g., at high flow or stirring rate, or at primary ion activities close to the lower detection limit of the measuring electrode where the exchange current densities are decreased significantly.[2,64,65]

The latter condition was fulfilled in dynamic response studies carried out in the range of the detection limit of the sensor.[152,153] For these experiments, an iodide-selective electrode of almost ideal Nernstian response and a lower detection limit of about 10^{-7} *M* iodide had been selected. Response time curves were recorded for activity increases and decreases, and the relevant measurements are surveyed in Chapter 2, Figure 20. In the first series of experiments (I), the activity of the conditioning solution ($a_{i,1}$) was changed by orders of magnitude, while the activity ratio (b) (Chapter 2, Equation 41) was maintained constant. In the second series of measurements (II), however, the value of b was varied from 10^{-1}, 10^{-2}, 10^{-3}, 10^{-4}, 10^{-5}, 10^{-6} at dilution, while it was increased from 10, 10^2, 10^3, 10^4, 10^5, 10^6 at concentration, and accordingly either the value of $a_{i,1}$ or $a_{i,2}$ was kept constant. The transient signals recorded at different activities were compared on the basis of normalized slope values (m_{eff}) (see Chapter 5, Section II).

In the first series of measurements (I), two different flow rates (v_1 = 140 mℓ/min, i.e., $v_1^{lin} \approx 3$ m/sec; v_2 = 100 mℓ/min, $v_2^{lin} \approx 2.1$ m/sec) were used, and the normalized slope values* were plotted as a function of the logarithm of the stepped activity value ($a_{i,2}$) (Figure 14).

The m_{eff} vs. log a_2 functions obtained by evaluating the transient signals corresponding to an activity increase (Figure 14a) or decrease (Figure 14b) show almost the same character.

* $m_{eff} = m\,\dfrac{59.16}{E}$, where m is the experimentally determined slope of the response time curve, and E is the potential difference corresponding to the given activity change. The evaluation of m is discussed in Chapter 5, Section I.

The only important difference is that the sections parallel to the x axis are at higher values for activity increase than for activity decrease, since the slopes are steeper in the former case at a given flow rate, i.e., the response time is shorter for activity increase compared to activity decrease.

Furthermore, in Figure 14 it is shown that, above a critical activity level (a_R), the normalized slope values (which generally can be related to response time data), were found independent of activity, $a_{i,2}$ (x axis, parallel section). In a similar series of measurements, Johansson and Norberg[140] and later Denks[161] obtained very similar results by plotting the time constant of fitted exponential functions vs. $a_{i,2}$. However, this activity range was found to vary with the flow rate; i.e., at high flow rates, the sections parallel to the abscissa were smaller compared to those recorded at low flow rates.

It is also apparent from Figure 14 that, below a critical primary ion activity, the response time values increase and, in accordance with this, the slope values of the transient signals decrease with decreasing primary ion activity. In this range, the slope values of the transient signals are flow rate independent. This is in contrast with the expectation based on the diffusion model discussed above.

This can be demonstrated by taking as an example the data shown in Figure 14b recorded at primary ion activity decrease. The slope values of the transient signals recorded decrease when the value of the primary ion, $a_{i,2}$, is lower than 3.10^{-5} *M* at a flow rate of 100 mℓ/min or lower than 10^{-3} *M* at a flow rate of 140 mℓ/min, while above these activity values, the slope of the transient response curves are found to be constant within the experimental error.

On the basis of these experimental findings, the contradictory literature data on concentration level-independent[5,91,125,127,139,154] as well as concentration level-dependent[4,17,18,124,134,136,140,152,153,155-162] response time data can be understood.

The experimental results shown in Figure 14 can be explained only partly with the film diffusion model (Figure 2) because, according to this model, the slope values of the response time curves at time t (e.g., initial slope values) are expected to be independent of the activity values in the whole activity range (Equations 37 to 39):[8,228]

$$\left(\frac{dE}{dt}\right)_t = S^x\left(\frac{d \ln a_i}{dt}\right)_t = S^x\left(\frac{1}{a_i}\cdot\frac{d a'_i}{dt}\right)_t \tag{37}$$

where S^x is $\frac{RT}{ziF}$ mV.

By inserting the surface activities of Equation 17a or Equation 15 into Equation 37, one gets for the slope values the following equations, respectively: at relatively long times ($\sqrt{D't} \gg \delta$):

$$\left(\frac{dE}{dt}\right)_{t>>0} = S^x \frac{(a_{i,2} - a_{i,1})\dfrac{1}{\tau'}e^{-t/\tau'}}{a_{i,1} + (a_{i,2} - a_{i,1})(1 - e^{-t/\tau'})} \tag{38}$$

while at relatively short times ($\sqrt{D't} \ll \delta$):

$$\left(\frac{dE}{dt}\right)_{t\sim 0} = S^x \frac{(a_{i,2} - a_{i,1})\dfrac{d\,\mathrm{erfc}}{dt}\left(\dfrac{\delta}{2\sqrt{D't}}\right)}{a_{i,1} + 2(a_{i,2} - a_{i,1})\,\mathrm{erfc}\left(\dfrac{\delta}{2\sqrt{D't}}\right)} \tag{39}$$

Since $a_{i,2}$ is proportional to $a_{i,1}$ (Equation 41), $ba_{i,1}$ can be inserted instead of $a_{i,2}$ into Equations 38 and 39:

$$\left(\frac{dE}{dt}\right)_{t>>0} = S^x \frac{a_{i,1}(b - 1)\frac{1}{\tau'} e^{-t/\tau'}}{a_{i,1}[1 + (b - 1)(1 - e^{-t/\tau'})]} \tag{40}$$

$$\left(\frac{dE}{dt}\right)_{t\sim0} = S^x \frac{2\, a_{i,1}(b - 1)\frac{d}{dt}\,\mathrm{erf\,c}\left(\frac{\delta}{2\sqrt{D't}}\right)}{a_{i,1}\left[1 + 2(b - 1)\,\mathrm{erf\,c}\left(\frac{\delta}{2\sqrt{D't}}\right)\right]} \tag{41}$$

Thus, it is apparent that, according to the diffusion model, dE/dt should be independent of the primary ion concentration and it is determined by physicochemical constants (e.g., D') and experimental conditions (b,δ) as well as by time (t). As dE/dt is a function of δ, the determined m_{eff} values were found flow rate dependent. Moreover, their actual values as expected depend on the direction of the activity change (cf. Figure 14a and b), since $b \ll 1$ for activity decrease and $b \gg 1$ for activity increase.

On the other hand, the response time values were found to depend on the primary ion activity outside the range of validity of the diffusion model, i.e., below a critical activity (a_R).

The unexpected alteration in m_{eff} values below the critical activity value may be understood by assuming that processes other than diffusion (e.g., dissolution, crystallization, charge transfer processes)[24,25,136] become rate controlling in highly diluted solutions. This assumption also explains the fact that, below a given activity (a_x) (see Figure 14), an increase in flow rate does not result in an increase in m_{eff}. Thus, it is expected that at high diffusion effectivity (i.e., at thin diffusion layers), processes other than diffusion are manifested on the response time curves at higher concentrations.[6,146,160] This assumption is in complete agreement with the experimental findings shown in Figure 14, i.e., the higher the flow rate, the higher the activity at which the breakpoint on the m_{eff} vs. log ($a_{i,2}$) plot appears.

2. *The Effect of the Activity Ratios*

In the second series of measurements (II) shown in Chapter 2, Figure 20, the effect of activity ratios on the response time curves either at a constant level of $a_{i,1}$ or $a_{i,2}$ has been studied. Two limiting cases can be distinguished:

1. $b \gg 1$ when the activity is increased.
2. $b \ll 1$ when the activity is decreased.

In (1), Equation 40, reduces to

$$\left(\frac{dE}{dt}\right)_{t>>0} = S^x \frac{b\frac{1}{\tau'} e^{-t/\tau'}}{1 + b(1 - e^{-t/\tau'})} \tag{42}$$

while Equation 41 reduces to

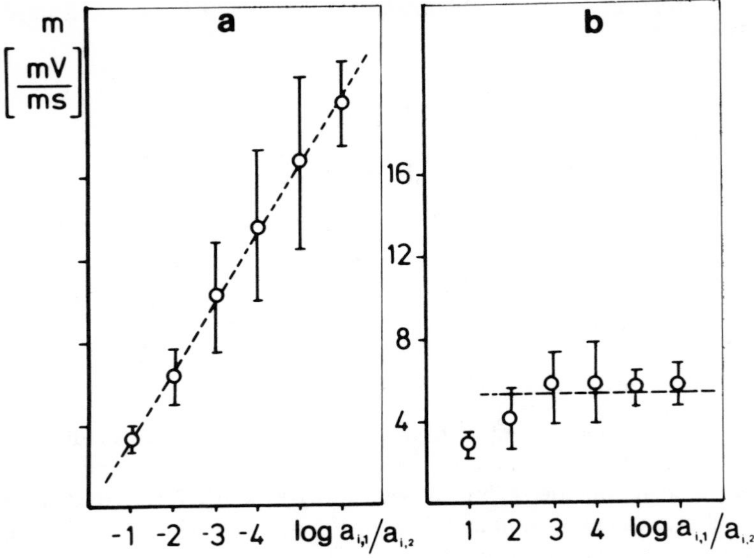

FIGURE 15. Initial slope values as a function of log $a_{i,1}/a_{i,2}$ in the second series of measurements (II in Chapter 2, Figure 20): (a) in the direction of concentration, $a_{i,2}$ = constant; (b) in the direction of dilution, $a_{i,1}$ = constant. The measured data are given along with their standard deviation. Flow rate was 140 mℓ/min.

$$\left(\frac{dE}{dt}\right)_{t\sim 0} = S^x \frac{2\,b\,A}{1 + 2\,b\,\text{erf c}\left(\dfrac{\delta}{2\sqrt{D't}}\right)} \tag{43}$$

where $A = \dfrac{d}{dt}\,\text{erfc}\,(\delta/2\sqrt{D't})$.

It follows from Equations 42 and 43 that dE/dt is proportional to b when the second term in the denominator can be neglected compared to unity. However, if the second term is much larger than unity, dE/dt becomes independent of the activity ratios. The former case is encountered when the initial slope of the response time curves is determined as

$$\lim_{t\to 0}\,(1 - e^{-\frac{t}{\tau'}}) = 0 \tag{44}$$

and

$$\lim_{t\to 0}\,\text{erfc}(\delta/2\sqrt{D't}) = \text{erfc}\,\infty = 0 \tag{45}$$

while the latter holds when the slope value is determined at the asymptotic part of the dynamic response curves.

The latter was experimentally proved by Uemasu and Umezawa[163] who determined the differential quotient ($\Delta E/\Delta t$) of the response time curves of copper (II) ISE at the asymptotic section of the curves. The response time data (the time needed to reach a preliminary defined slope value; see Chapter 5.2) were found to be independent of the activity ratios [Equation 42; $1 \ll b\,(1 - e^{-t/\tau'})$].

On the other hand, Lindner, et al.[152,153,165] showed (Figure 15) that the initial slope values

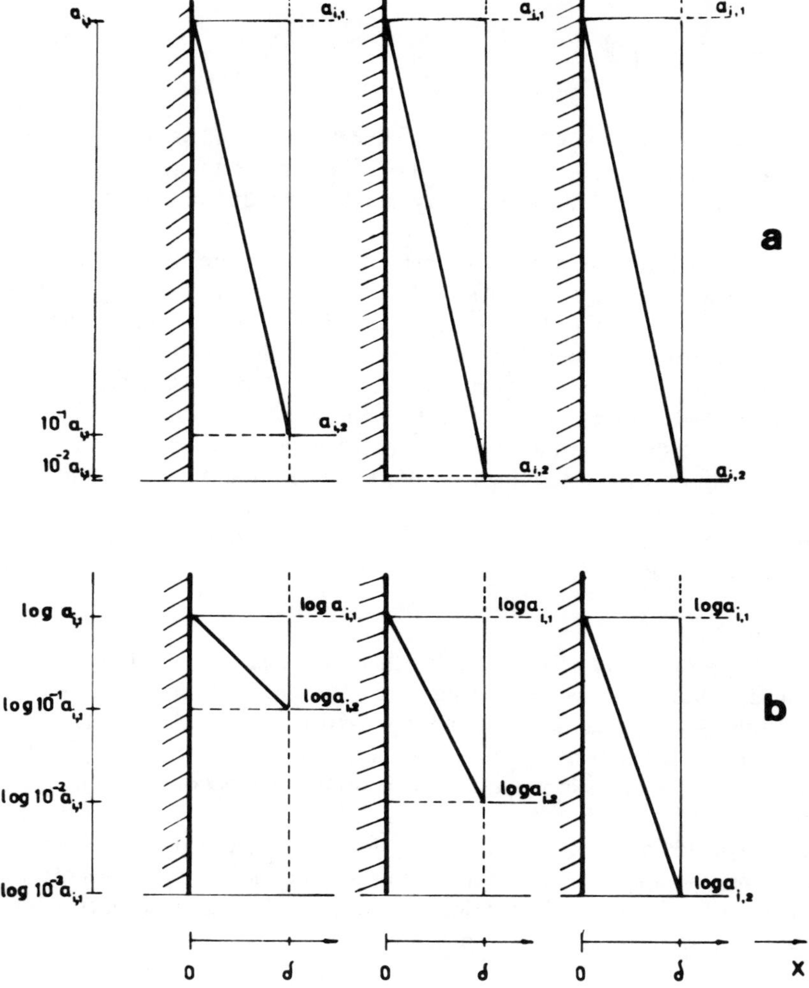

FIGURE 16. The activity (a) and log activity (b) gradients in the adhering film at t = 0, the second series of measurements (II) in Chapter 2, Figure 20.

increased proportionally with log b = log $a_{i,1}/a_{i,2}$ instead of b [Equation 42, $1 \gg b\,(1 - e^{-t/\tau'})$].

Both experimental results [case (1)] can formally be interpreted by considering an almost constant concentration gradient within the adhering layer at different activity ratios (Figure 16a). Thus, in a series of experiments (II) in Chapter 2, Figure 20, when $a_{i,2} = 10^{-1}$ *M* and activity steps are applied starting from different activity levels (e.g., $10^{-6} \rightarrow 10^{-1}$, $10^{-5} \rightarrow 10^{-1}$, $10^{-4} \rightarrow 10^{-1}$), then both the surface activity (a_i') and the concentration gradient within the adhering layer attain an almost identical value after a short initial period (t′) independently from the initial activity, $a_{i,1}$.

In the initial period ($0 < t < t'$), the slope of the transient signals should be proportional to $\Delta E = E(t') - E(a_{i,1}) \approx \log (a_{i,2}/a_{i,1})$, i.e., to log b.

In case (2), when an activity decrease is introduced, $b \ll 1$, Equation 40 is simplified to

$$\left(\frac{dE}{dt}\right)_{t>>0} \approx \frac{S^x}{\tau'} \tag{46}$$

and Equation 41 becomes

$$\left(\frac{dE}{dt}\right)_{t\sim 0} = -2S^{x} \frac{d/dt\ \mathrm{erfc}(\delta/2\sqrt{D't})}{1 - 2\ \mathrm{erfc}(\delta/2\sqrt{D't})} \tag{47}$$

From both Equations 46 and 47, one would expect activity independent slope values. In agreement with the expectations, Lindner et al.[152,153] reported on fairly constant experimentally determined initial slope values at activity increase with different activity ratios (Figure 15b).

In connection with the above, it should be mentioned that the agreement between theoretical expectations drawn from the diffusion model and experimental results highly depends on the dynamic characteristics of the ISE and cell design. In other words, the agreement depends on how far the experimental conditions are met with model assumptions.

It is also apparent from the foregoing considerations that the shape of the m vs. log ($a_{i,1}/a_{i,2}$) plots depends to a large extent on the time instant of differentiation. Obviously, the large variation of the experimentally determined slope values can also be attributed to the fact that the slopes were determined at different time instants. In addition to the above discrepancy, other sources of error must also be taken into consideration, namely, the insufficient reproducibility of the experimental conditions, and inaccurate graphical evaluation of the slope values (see Chapter 5, Section II).

3. Conclusions

The foregoing results led to the conclusion that ion transport through the hydrodynamic boundary layer is often the rate-determining step of the electrode process of ISE of constant composition (glass-, precipitate-, and liquid ion-exchanger-based electrodes). The relationships derived on the basis of the model shown in Figure 2 (Equation 19) permit the interpretation of many characteristic properties of the transient functions and the mathematical description of the latter.

On the other hand, by applying sufficiently high flow-rate and/or by working in very diluted solutions, the transient functions are primarily defined by processes other than diffusion. Under these experimental conditions, the shape of the transient functions is affected by the flow rate to a decreasing extent or not at all.

B. The Multielectrode Model

The thickness of the hydrodynamic boundary layer is determined according to Equations 28, 30, and 34, by the solvent properties (ν, D), the flow velocity, as well as the geometry of the measuring setup (1, x, d_n).

Under the measuring conditions of our experimental device (Chapter 2, Figure 16; Figure 13),[4,5,6,8,151-154] the diffusion layer thickness increases outward from the center of the electrode surface if the solution beam is placed centrally. In other instruments[144-147], the thickness of the boundary layer increases in the direction of flow, e.g., along the surface if the solution is streamed parallel to the surface. Accordingly, the components of the stepped solution $a_{i,2}$ have to pass through the increasing diffusion layer. Thus, different points of the electrode surface are in contact with solutions of different activity ($a_i' = f(t,1)$ and, therefore, different response time values are expected at different points of the electrode surface. The measured potential is the "average" of these values. Consequently, the electrode can be regarded as a multielectrode system, composed of many elementary electrodes. According to this definition, the elementary electrode is defined as a part of the macroelectrode on the surface of which a homogeneous concentration distribution exits. Thus, the measured potential is the sum of potentials of the elementary electrodes, which can be termed mixed potential.[6,14]

Thus, the response characteristics of ISE of relatively large surface area can be understood by considering the time dependence of the mixed potential of the multielectrode system.

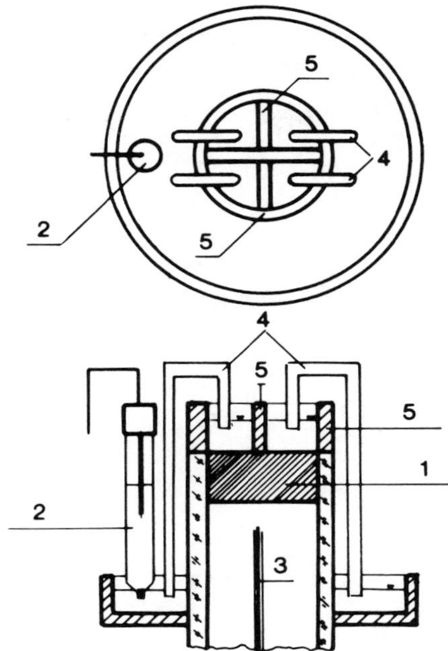

FIGURE 17. Measuring setup for modeling the nonhomogeneous concentration distribution at the boundary of an electrode of relatively large surface area. (1) Ion-selective membrane; (2) outer reference electrode; (3) inner reference electrode; (4) salt bridge; and (5) wall for separating the different sections of the electrode surface.

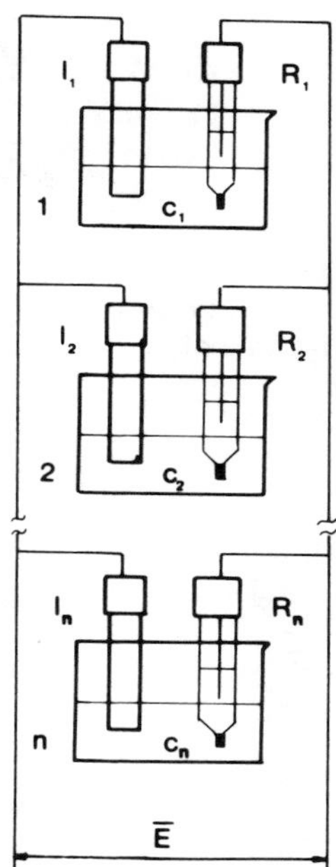

FIGURE 18. Measuring setup for modeling the nonhomogeneous concentration distribution at the boundary of an electrode of relatively large surface area. I = indicator electrode; R = reference electrode; c = concentration of the sample solution; and 1,2 . . . n = indexes referring to the number of the cell.

Two experimental setups (shown in Figure 17 and 18) were developed for modeling the mixed potential of ion-selective membranes in contact with solutions of different concentration. In the one shown in Figure 17, the relatively large surface area (ø = 12 mm) of a specially designed electrode was divided into sections of equal surface areas and connected by separate salt bridges to the same reference electrode.

Solutions of different primary ion activities were placed in each section and the mixed potential was measured. In the setup shown in Figure 18, the emf values of cells incorporating different types of electrodes and solutions of different concentrations were measured separately and after connecting the individual cells in parallel to each other.

With both setups, a dynamic system is approximated with a static model and, therefore, provides information corresponding to one point of the response time curve. Thus, the model provides only qualitative information for the dynamic system.

The experimental results shown in Tables 3 and 4 were evaluated on the same basis as used for metallic electrodes in electrode kinetics.

Thus, the ion-selective membrane was assumed to behave similarly to a silver metal electrode of negligibly small resistance as far as the mixed potentials measured in experimental setups shown in Figures 17 and 18 are concerned. (With metal electrodes of low resistances, one can consider the metal electrode an equipotential system.)

If the electrode surface is divided into parts (Figure 17) and the individual segments are bought in contact with solution containing primary ions in different concentrations, then the relevant current densities in the corresponding segments are expected to be different. As an example, let us consider when the electrode surface is divided into two parts:

$$F_1 = \beta F \tag{48a}$$

Table 3
COMPARISON OF THE "APPARENT" IONIC ACTIVITIES CALCULATED BY EQUATION 53 AND FROM THE POTENTIAL VALUES MEASURED IN THE CELL SHOWN IN FIGURE 17

Surfaces covered with solutions of different activities	Measured apparent activities	Calculated apparent activies (Equ. 3.53)
	1.22	1.15
	1.65	1.4
	1.33	1.17
	2.01	1.46
	2.24	2.26
	2.4	2.49
	206	2.11

10^{-1} M 10^{-2} M 10^{-3} M

$$F_2 = (1 - \beta)\,F \tag{48b}$$

where F is the geometric surface area; F_1 and F_2 are the surface area of the individual segments; and β is the proportionality factor defined by Equation 48a. Supposing that F_1 is in contact with solution $c_{o,1}$ while F_2 is with $c_{o,2}$, the corresponding current densities can be given on the basis of Equations 22 and 25 in Chapter 2 as follows. For surface area F_1:

$$j_1 = j_a - j_{c,1} = k'_a c_R - k'_c c_{o,1} \tag{49a}$$

while for surface area F_2:

$$j_2 = j_a - j_{c,2} = k'_a c_R - k'_c c_{o,2} \tag{49b}$$

where c_R and c_o are the concentrations of the reduced and oxidized species and k'_a and k'_c can be given by comparing Equation 49a and Equation 25 in Chapter 2:

Table 4
CALCULATED AND MEASURED EMF VALUES OF CELLS SHOWN IN FIGURE 18 WITH INDICATOR ELECTRODES OF DIFFERENT RESISTANCES

No. of cells	Solutions in the cells, $AgNO_3$ (*M*)	Electrode couples	Measured cell voltage (mV)	Data obtained after coupling cells 1 and 2 in parallel		
				Measured cell voltage	Measured apparent activities	Calculated apparent activities (Equation 53)
1	10^{-3}	Ag_{met}-SCE	373	371	3.03	3.26
2	10^{-4}	AgI_{membr}-SCE	314			
1	10^{-3}	AgI_{membr}-SCE	373			
2	10^{-4}	Ag_{met}-SCE	314	340	3.56	3.26

$$k'_a = k_a \cdot \exp \frac{\alpha F \epsilon}{RT} \tag{50a}$$

$$k'_c = k_c \cdot \exp\left(- \frac{(1 - \alpha) F\epsilon}{RT} \right) \tag{50b}$$

Under potentiometric conditions, no external current flows in the galvanic cell. Thus,

$$F_1 i_1 + F_2 i_2 = 0 \tag{51}$$

$$\beta \cdot (k_a c_R - k_k c_{o,1}) + (1 - \beta) \cdot (k_a c_R - k_k c_{o,2}) = 0 \tag{52}$$

and $\bar{c}_o$, i.e., the apparent concentration of the potential determining ion is

$$\bar{c}_o = \frac{k'_a}{k'_c} \cdot c_R = \beta c_{o,1} + (1 - \beta)\, c_{o,2} \tag{53}$$

According to the former kinetic description, a metal electrode being in simultaneous contact with solutions of different concentration acts as if its surface were entirely covered with a solution of concentration, $\bar{c}_o$, expressed by Equation 53.

Although most of the experiments were performed with ISE of relatively high resistance, a surprisingly good agreement was found between apparent concentrations, $\bar{c}_o$, calculated with Equation 53, and those derived from cell potential obtained with ISE (Table 3).[8]

The impressive good correlation found between the apparent activities measured and calculated (Table 3) in the case of ion-selective membranes of high resistance supports the earlier findings[13,14,65,66] that the kinetics developed for metallic electrodes are also applicable for ISE.

Based on the experimental results and model calculations, it is evident that the mixed potential values are determined mainly by the primary ion activity of the cell containing the most concentrated solution. The mixed potential of ISE could be calculated more precisely if the resistances of the individual membrane sections were known.

The pronounced effect of the electrode resistance on the mixed potentials were verified by the experiments shown in Figure 18. In one of the parallelly connected cells, a metallic silver indicator electrode was used, while in the other, a silver iodide-based ion-selective membrane electrode served the same purpose. The geometric surfaces of both electrodes were approximately equal. The corresponding mixed potential data are found in Table 4, which shows the trend that the mixed potential values were closer to those of the silver electrode than expected on the basis of calculations performed using Equation 53. It means that the potential of an electrode of lower resistance plays a more definite effect on the formation of the mixed potential.

In addition, the mixed potentials were slightly affected by the solution resistance. In fact, the difference between calculated and measured data was found to be smaller if the experiments were carried out in solutions of high ionic strength.

In the light of these results, the effect of the direction of concentration change on the transient functions of ISE or on the response time data, defined as one point of the letter, can be explained from a different aspect. Namely, the response time is expected and found to be short at a concentration increase since it is known that the mixed potential is determined primarily by the presence of the more concentrated solution even if the latter is present in only a fraction of the electrode surface. On the contrary, the response time is expected to be long at concentration decrease because the more concentrated solution remaining on the

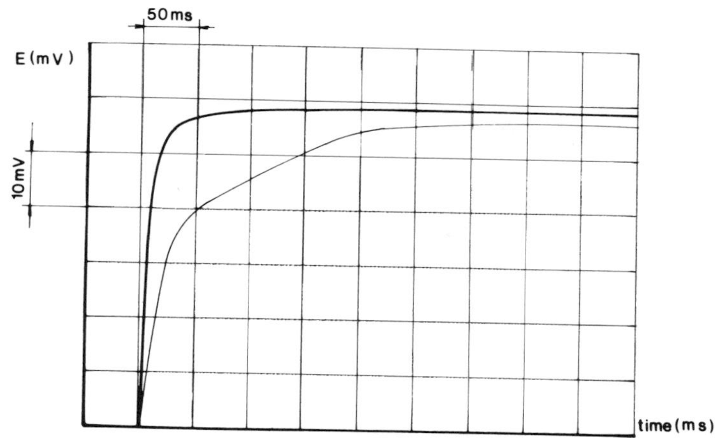

FIGURE 19. Effect of electrode surface area on the response time curve of a silver-selective electrode (electrode membrane 1:1 Ag_2S/AgI): v = 117 mℓ/min, $a_{i,1} = 10^{-3}$ *M* $AgNO_3$, $a_{i,2} = 10^{-2}$ *M* $AgNO_3$. (lighter line) ø = 12 mm; (darker line) ø = 6 mm.

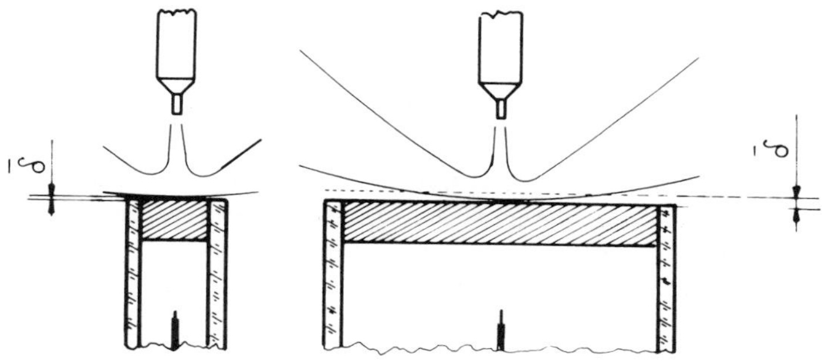

FIGURE 20. Effect of the geometric surface of the ISE area on the thickness of the hydrodynamic boundary layer compared to jet orifice. $\bar{\delta}$ is the average thickness of hydrodynamic boundary layer.

electrode surface after the activity step determines the potential even when a considerable part of the electrode surface is already in contact with the more diluted solution.

From the above, it is clear why the roughness of the membrane surface and the presence of scratches and cavities have a more pronounced effect on the dynamic response characteristics of the sensor, especially at concentration decrease.

The multielectrode model based on the assumption of inhomogeneous concentration distribution at the electrode surface permits a straightforward interpretation of the effect of the electrode surface on the transient functions. The fact that a decrease in the electrode surface area results in shorter response time (Figure 19) proves that ion diffusion within the hydrodynamic boundary layer is rate determining.

Consequently, the decrease of the electrode surface area compared to the jet orifice results in the decrease of the average layer thickness (Figure 20; Equations 28 and 30), which means that ion transport is faster. This is advantageous in order to (1) study electrode processes following film diffusion, and (2) design flow-through potentiometric cells of short response time. In other words, the disturbing effect of inhomogeneous concentration distribution can be eliminated by using point-like electrodes placed as close as possible to the outlet of the jet and by increasing the flow velocity.

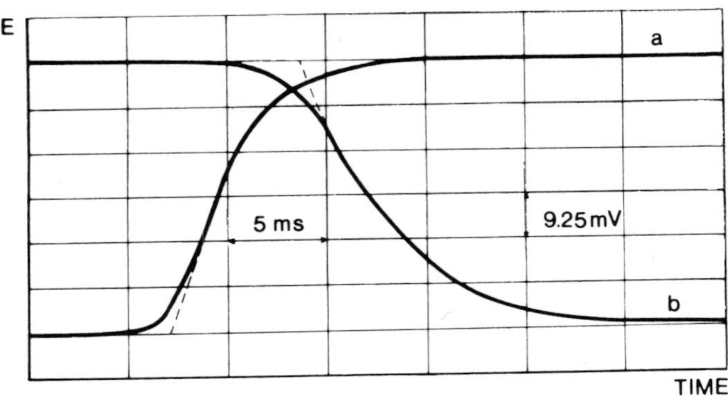

FIGURE 21. The transient signals of a silver metal electrode recorded with high resolution in respect of time recorded in the measuring set-up of Denks and Neeb (see Chapter 2, Figure 18). Activity step: $10^{-1} \rightleftharpoons 10^{-2}$ *M* Ag^+; ionic strength: 0.1 *M* ($NaNO_3$). (a) Activity increase; (b) activity decrease. (From Denks, A., Vergleichende Untersuchungen über das Ansprechverhalten von metallischen Ag-Elektroden und Ag_2S/AgX (X = Cl, I). Festkörpermembranelektróden bei schnellen Konzentrationsänderungen in strömenden Lösungen, Dissertation, Johannes Gutenberg-Universität, Mainz, 1977. With permission.)

It was found experimentally that the transient functions of sliver iodide-based electrodes of different surface areas were practically identical (cf. Chapter 2, Figure 21) at high flow velocities (v > 100 mℓ/min). This phenomenon may be explained with the help of Equation 30, describing the thickness of the hydrodynamic boundary layer at turbulent flow profile, as in this expression the characteristic length "1" is at power 0.1, or by assuming that the processes following diffusion increasingly become the rate-determining ones in the overall electrode process at this relatively high flow rate.

In contrast to this, the transient functions of electrodes with large surface area at low flow rates are often composed of several sections (Figure 19). This means that under these conditions, the flow profile especially at the edges is ill-defined and remixing cannot be avoided especially in wall-jet arrangements.

III. KINETICS OF INTERFACIAL REACTIONS

A. Energy Barrier Concept

As it has been pointed out in the foregoing sections, under theoretically predicted measuring conditions (see Figure 14), the activity step method can provide information about the electrode processes following diffusion through the stagnant solution layer. Rechnitz and Hameka[206] in studying the potential-time functions of glass electrodes after a step change in pH assumed that the rate determining step among the partial reactions of the overall electrode response was the charge transfer across the electrical double layer.

For describing the transient potential functions of glass electrodes Rechnitz and Hameka[206] as well as Johansson and Norberg[140] used in principle an energy barrier model in which, besides surface resistance, double layer capacitance also plays an important role. According to the energy barrier model, the equilibrium at the membrane solution interface, characterized by the exchange current density, is upset by the activity step, i.e., by chemical potential change of the solution. In order to restore the equilibrium, current flows across the phase boundary (energy barrier) towards the direction of decreasing chemical potentials. The potential time function thus can be given by the simple relationship corresponding to the discharge of a condenser (i.e., the double layer capacitance).

Based on this model, Rechnitz and Hameka[206] derived the following exponential-type relationship:

$$E(t) = E_1 + (E_2 - E_1)(1 - e^{-kt}) \tag{54}$$

where k is a constant characteristic to the charge transfer process, which depends on the final concentration of the activity step, temperature, etc. According to this model, the shape of the transient functions depends only on the stepped activity value, which means that if the final activity value is the same, then the shape of the transient signals is independent of the direction of the concentration change.

In view of the similarity of the forms of Equations 21 and 54, the latter is expected to be valid only in the so-called linear regime.

Johansson and Norberg[140] further developed the energy barrier concept suggested by Rechnitz and Hameka,[206] also taking into consideration the direction of the charge transfer reaction.

Shatkay[134] critically assessed the energy barrier models, as well as the validity of the model assumptions and the consequences of the simplifications made. Accordingly, the corrected form of the equation of Johansson and Norberg is the following:[134]

$$E(t) - E_1 = (E_2 - E_1) - \frac{2\,RT}{F} \ln \frac{1 + e^N \cdot e^{-kt}}{1 - e^N \cdot e^{-kt}} \tag{55}$$

where

$$e^N = \frac{\sqrt{\dfrac{a_2}{a_1}} - 1}{\sqrt{\dfrac{a_2}{a_1}} + 1} \tag{56}$$

Equation 55 is exact (provided that the model assumptions can be accepted) but difficult to handle. Therefore, Johansson and Norberg,[140] noting that the exponential term of Equation 55 is small compared to unity, substituted the logarithmic term by the approximate expression: $\ln(1 + x)/(1 - x) \approx 2\,x$. Thus, the simplified equation obtained was formally the same as that of Rechnitz and Hameka (Equation 54), but the significance of the constant k is different, being besides others, a function of both $a_{i,2}$ and $a_{i,1}$.

Denks[161] derived Equation 55 in a simpler way by considering that the final value of the overpotential corresponding to the current imposed by the galvanostatic current step is not attained instantaneously becuase of the charging of the double layer capacitance:

$$j = j_{CD} + j(\eta_t) \tag{57}$$

where j is the galvanostatic polarization current density; $j(\eta_t)$ is the charge transfer current density; and j_{CD} is the charging current density of the double layer capacity, C_D:

$$j_{CD} = C_D \cdot \frac{d\eta_t(t)}{dt} \tag{58}$$

where η_t is the charge transfer overpotential at short times: $(t \rightarrow 0)$.

Denks[161] assumed that j_{CD} can be substituted as expressed by the Buttler-Volmer equation (Chapter 2, Equation 23), and obtained the following differential equation:

$$\frac{d\eta_t(t)}{dt} = -\frac{j_0}{C_D}\left(\exp\frac{\alpha zF\eta_t(t)}{RT} - \exp\frac{-(1-\alpha)\, zF\eta_t(t)}{RT}\right) \qquad (59)$$

If $\alpha = 0.5$ and $z = 1$, then Equation 59 can be simplified:

$$\frac{d\eta_t(t)}{dt} = -\frac{j_0}{C_D}(e^{A\eta_t(t)} - e^{-A\eta_t(t)}) \qquad (60)$$

where

$$A = \frac{F}{2\,RT}$$

Furthermore, Denks supposed that the transient function of ISE is determined by only charge transfer overpotential. To solve Equation 60, the boundary conditions for the integration were as follows:

$$t=0 \quad \eta_t(t = 0) = E_2 - E_1$$

$$t \quad \eta_t(t) = E_2 - E_t$$

$$t=\infty \quad \eta_t(t = \infty) = 0$$

which led to Equation 55.

Summing up, it can be stated that the exponential relationships derived on the basis of the energy barrier model are applicable first of all when small activity steps are applied (i.e., in the linear regime). Moreover, before applying these models, the system must be tested carefully to determine whether the charge transfer process across the electrical double layer is really the rate-determining one. On the contrary, faulty conclusions may be drawn, in spite of the fact that these relationships (Equations 54 and 55) can be fitted to experimental transient functions of potentiometric cells (Figure 5) due to their similarity to equations derived under different model assumptions (see Equations 19 and 21; Section III.B).

B. First-Order Chemical Kinetics and the Consecutive Reaction Model

Tóth and Pungor[4,5,104,154] reported first on the transient functions of nonglass ISE using specially designed equipment. The transient functions were interpreted on the basis of a different model assumption, i.e., first-order kinetics. Accordingly, as a rate-determining reaction step, the ion-dehydration (desolvation) was supposed. The desolvation reaction is

$$IS_x \overset{k}{\rightleftharpoons} I + xS \qquad (61)$$

where I is the potential determining ion (the charge is omitted for the sake of simplicity); S is a solvent molecule; x is the solvation number; and k is the rate constant of the dehydration (desolvation) reaction. The rate of that reaction is

$$\frac{d\,a'_i(t)}{dt} = -k\, a_{IS_x} t \qquad (62)$$

where a'_i is the activity of desolvated ion I at the electrode surface; a_{IS_x} (t) is the activity of solvated ions at phase boundary following an activity step from $a_{i,1}$ to $a_{i,2}$; and

$$a_{IS_x}(t) = a_{i,2} - a'_i \tag{63}$$

The surface activity of ion I can be obtained by integrating Equation 62:

$$\int_{a_{i,1}}^{a'_i} \frac{da'_i}{a_{i,2} - a'_i} = -k \int_0^t dt \tag{64}$$

yielding:

$$a'_i = a_{i,2} - (a_{i,2} - a_{i,1})\, e^{-kt} \tag{65}$$

Inserting Equation 65 into the Nernst equation, the time dependence of the electrode potential is obtained:

$$E(t) = E_2 + S \log\left[1 - \left(1 - \frac{a_{i,1}}{a_{i,2}}\right) e^{-kt}\right] \tag{66}$$

This relationship has the same form as Equation 19 derived by assuming that diffusion across the hydrodynamic boundary layer is the rate-determining process. However, the physicochemical significance of the constants in the exponent of the two equations is different.

The effect of the direction of the concentration change on the transient functions can be interpreted on the basis of the concentration dependence of the reaction assumed to be the rate-determining one (cf. Equation 62). The direction of concentration change also determines the factor $(1 - a_{i,1}/a_{i,2})$ in both Equations 66 and 19.

It has been noted in connection with the generalization of the model based on reaction kinetics[6] that several physicochemical models can formally be associated with first-order kinetics. Thus, in addition to dehydration, either a chemical reaction preceding the electrode reaction, or adsorption (chemisorption) of ions, or the ion-exchange process, etc., can also determine the reaction rate. However, if the rate-determining step of the electrode process only affects the constant term in the exponent of the equations of the transient function, neither of these assumptions can be verified unless the physicochemical constants are known.

The general model based on reaction kinetics also permits the assumption that the component determining the electrode potential is formed in a higher-order reaction or in consecutive reactions of commensurable rate constant.[8]

The simplest consecutive overall reaction consists of two partial ones, both following first-order kinetics:

$$A \xrightarrow{k_1} B \xrightarrow{k_2} C \tag{67}$$

where k_1 and k_2 are the corresponding first-order reaction rate constants.

Naturally, in addition to chemical processes transport and other physical processes may also be included in a consecutive reaction series. Therefore, in a broader sense, one may handle all of these consecutive reactions mathematically in the same way if they can be treated as a first approximation as reactions of the same order (cf. Equations 17a and 65).

In the case of the reaction scheme given by Equation 67, the surface activity becomes[228,229]

$$a'_i = a_{i,2}\left[1 - \left(1 - \frac{a_{i,1}}{a_{i,2}}\right) \frac{k_2 \cdot e^{-k_1 t} - k_1 \cdot e^{-k_2 t}}{k_2 - k_1}\right] \tag{68}$$

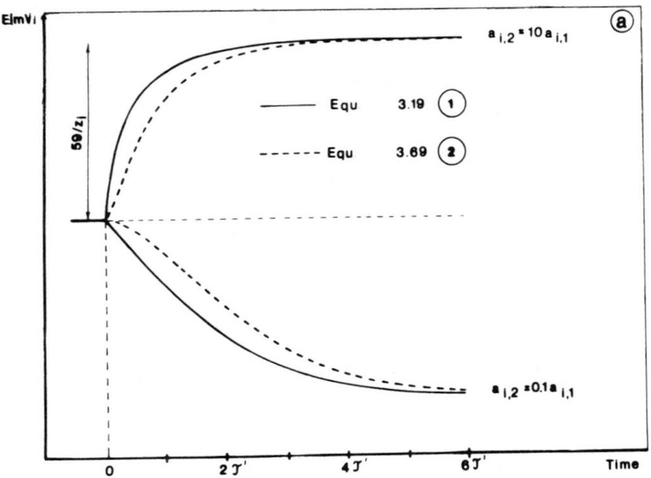

FIGURE 22. Theoretical potential response vs. time profiles calculated according to the film diffusion (Equation 19; curve 1) and the consecutive reaction kinetic model (Equation 69; curve 2). In (b), the curve corresponding to Equation 69 is shifted parallel for a better comparison.

If the rate constants of the two processes differ to a large extent, i.e., $k_1 \gg k_2$ or $k_1 \ll k_2$, Equation 68 can be simplified to yield Equation 65 (cf. Equation 17). The substitution of Equation 68 into the Nernst equation yields the potential vs. time relationship when the electroactive species is generated by coupled first order reactions:

$$E(t) = E_2 + S \log \left[1 - \left(1 - \frac{a_{i,1}}{a_{i,2}} \right) \frac{k_2 \cdot e^{-k_1 t} - k_1 \cdot e^{-k_2 t}}{k_2 - k_1} \right] \tag{69}$$

When studying the dynamic response curves of ISE at high time resolution (Figure 21), one may observe that the initial parts of these curves have an almost zero-slope value in contrast with theoretical expectation. These initial parts, in general, have been neglected in the extrapolation of the rising — almost linear section — of the curves to "t = 0". This contradiction (almost zero-slope section) was ascribed to the nonideality of the activity step.[6]*

However, by comparing the curves calculated by Equations 19 and 69, it can be seen clearly that the two curves run parallel except for a short initial period. Thus, by the extrapolation of the dynamic response curves to "t = 0", important information can be lost (Figure 22).

C. Second-Order Chemical Kinetics

The response time of ISE is considerably increased if the electrodes are used in very dilute solutions, i.e., in the range of the lower limit of detection (cf. Section II.A)[24,25,134,136,152,153,156,157]

The relationships (Equations 19a, 54, and 66) discussed in the previous section in many cases failed to be appropriate for the interpretation of the transient functions recorded in dilute solutions. However, hyperbolic functions were found to be applicable for describing the transient functions ranging from some minutes to some hours.

The empirical relationship suggested by Müller:[230]

* The short time behavior of different types of ISE can also be described mathematically in terms of the film diffusion model using the solution suggested for the time range $\sqrt{D't} \ll \delta$ (Equation 10a; Figure 3), too.

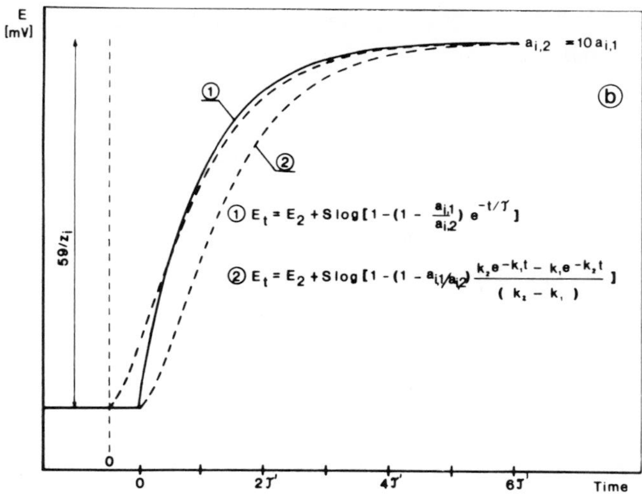

FIGURE 22b.

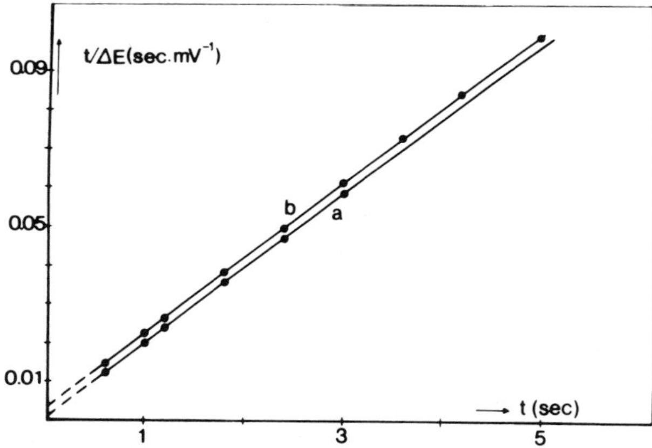

FIGURE 23. Comparison of Equation 70 with experimental data; (a) $1 \times 10^{-4} \rightarrow 1 \times 10^{-3}$ *M* F^-; (b) $1 \times 10^{-3} \rightarrow 1 \times 10^{-4}$ *M* F^-. (From Mertens, J., Van den Winkel, P., and Massart, D. L., *Anal. Chem.*, 48, 272, 1976. With permission.)

$$\Delta E(t) = \frac{t}{a + b \cdot t} \tag{70}$$

where $\Delta E(t)$ is the potential change due to a sudden increment in primary ion activity, and a,b are constants, yielded excellent results in the evaluation of the transient functions of silver sulfide-based silver ISE and lanthanum fluoride-based fluoride ISE.[141]

Equation 70 can be rewritten in the following form:

$$\frac{t}{\Delta E(t)} = a + bt \tag{71}$$

Thus, a plot of $t/\Delta E(t)$ vs. t yields a straight line (Figure 23) with a slope b and an intercept a. It is easy to show that the new equilibrium value E_2 is given by 1/b.

This equation is identical to that suggested by Shatkay[134] if $K = b/a$ and $b = 1/(E_2 - E_1)$, and was found to yield excellent fitting to the experimental curves of Savage and Isard:[123]

$$E(t) = E_1 + (E_2 - E_1)\frac{Kt}{1 + Kt} \tag{72}$$

where K is a constant.

Buffle et al.[24,25] also employed a hyperbolic function to describe the transient functions of fluoride and chloride electrodes in dilute solutions:

$$E(t) = E_2 - \frac{1}{At + B} \tag{73}$$

where A and B are empirical constants. Except for the omission of the constant term B, a similar empirical equation was suggested by Orion Research Inc.,[162] for estimating equilibrium potential E_2 of sodium-selective glass electrodes, which can also be transformed in Müller's equation (Equation 70) by introducing:

$$B = b = \frac{1}{E_2 - E_1}$$

and

$$A = \frac{b^2}{a}$$

Another transformation of Equation 73 was also suggested[24,25] for the evaluation of the equilibrium potential E_2:

$$\frac{E(t)\,t - E_i t_i}{E(t) - E_i} = E_2 \frac{t - t_i}{E(t) - E_j} - \frac{B}{A} \tag{74}$$

where E_i is the potential at time t_i (arbitrarily chosen, generally close to zero). A plot of the left-hand side of Equation 74 vs. $\frac{t - t_i}{E(t) - E_i}$ gave a straight line of slope E_2 (Figure 24). Since E_2 was found to be independent of the arbitrarily chosen value of t_i, E_2 could be determined from Equation 74 by recording a small portion of the $E = f(t)$ function.

Buffle et al.[24,25] proved experimentally that, in very dilute solutions, the dissolution of the electrode membrane is the rate-determining step in the electrode response (cf. Chapter 2, Section III and Figure 12). The E vs. t relationship for $t \geqslant 1$ min was found to be a hyperbolic function (Figure 24). The rate of growth or dissolution of crystals often follows second-order kinetics,[24,231,232] namely

$$\frac{d\Delta a_i(t)}{dt} = -\,k \cdot [\Delta a_i(t)]^2 \tag{75}$$

where $a_i(t)$ is the difference between the actual surface activity a'_i (sensed by the electrode) and the final equilibrium value $a_{i,2}$:

$$\Delta a_i(t) = |a'_i(t) - a_{i,2}| \tag{76}$$

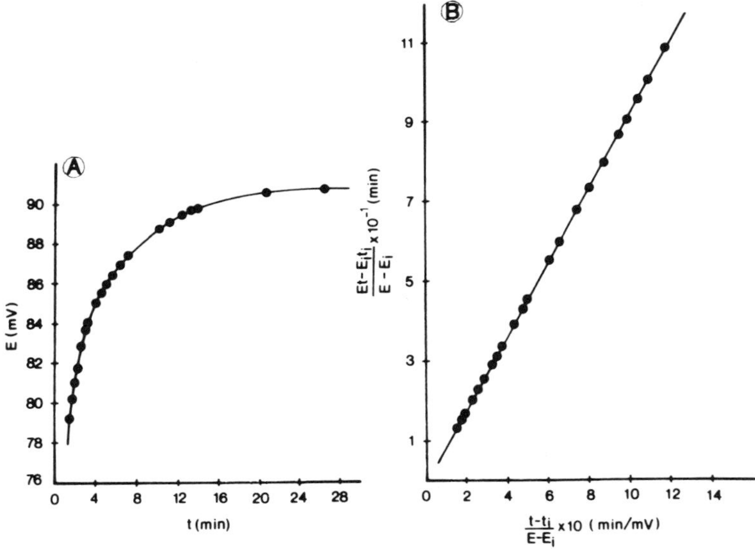

FIGURE 24. (A) E vs. time curve for a fluoride electrode (immersion technique). Concentration change from 10^{-2} to 10^{-4} *M* F^-, I (ionic strength) = 1 *M* $NaNO_3$, T = 25°C. (The electrode used was old. Similar curves are obtained for new electrodes except that the time scale is reduced.) (B) Verification of empirical relationship, with the experimental data of the curve given in (A). (From Buffle, J. and Parthasarathy, N., *Anal. Chim. Acta*, 93, 111, 1977. With permission.)

The integration of Equation 75 yields:

$$a'_i = a_{i,2} - (a_{i,2} - a_{i,1}) \frac{1}{1 + |a_{i,2} - a_{i,1}|\, kt} \tag{77}$$

Introducing surface activity a'_i into the Nernst equation, we obtain:

$$E(t) = E_2 + S \log \left\{ 1 + \frac{\dfrac{a_{i,1}}{a_{i,2}} - 1}{1 + a_{i,2}\left|1 - \dfrac{a_{i,1}}{a_{i,2}}\right| kt} \right\} \tag{78}$$

Equation 78 can be transformed into Equation 73 if the logarithmic term is linearized. Then the constants of Equation 73 are

$$A \cong \frac{F}{RT} \cdot ka_{i,2}; \; B \cong \frac{F}{RT} \frac{a_{i,2}}{|a_{i,2} - a_{i,1}|}$$

This model was experimentally verified by Buffle et al.[24,25] by showing the dependence of A and B/A parameters on the primary ion activity ($a_{i,2}$).

A hyperbolic function of the same form (Equation 70) was derived by Mertens et al.[141] on the basis of an equivalent circuit. However, the correspondence of the circuit elements and the processes determining the transient functions was not elucidated. On the basis of the flow rate dependence of the transient functions (cf. Section II and Chapter 5, Section V). Mertens et al. assumed that diffusion across the boundary layer is the rate-determining step in the electrode response.

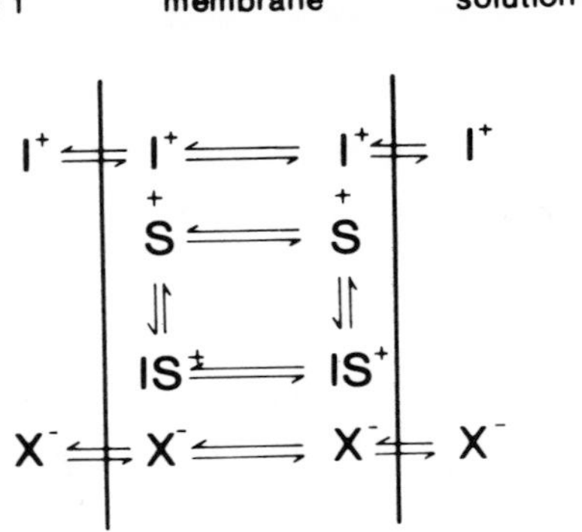

FIGURE 25. Schematic representation of a neutral carrier membrane system. (For analytically useful carrier membranes, the concentration of complexes IS_n^{z+} should exceed that of uncomplex cations I^{z+}, but should be smaller than the concentration of free ligands S.[17,132]

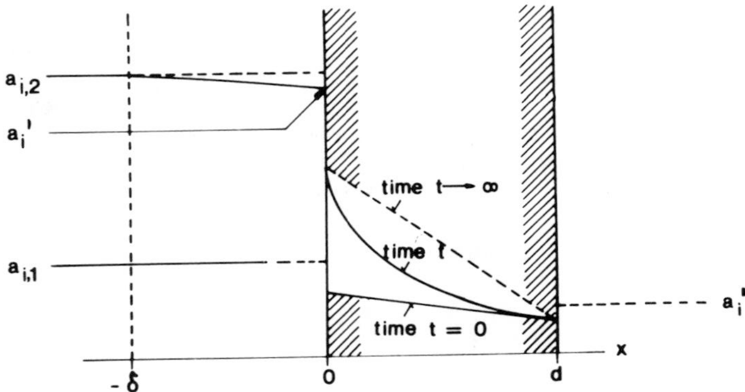

FIGURE 26. Diffusion model suited for neutral carrier membranes. The establishment of the steady-state concentration profile within the membrane for t > 0 is assumed to be rate controlling.

IV. DIFFUSION WITHIN THE ION-SENSING MEMBRANE

A certain increase in the respone time is expected if the transfer rate of charged carriers across the interface is controlled by diffusion processes within the membrane bulk.[7,17,132] In such cases, the transient functions (i.e., the response time) are expected to be affected by the composition of the membrane phase.[7,17,132,222]

The following treatment is based on the membrane system presented in Figure 25. It has been shown earlier that, in the analytically useful range[87,132] (in the range of low salt extraction), the concentration of cationic complexes at the membrane surface can be related directly to the outside activity of cations (Figure 26):[7,132]

$$\left.\begin{aligned} c_{IS}(0) &= K\, a'_i \\ c_{IS}(d) &= K\, a''_i \end{aligned}\right\} K << 1 \qquad (79)$$

where c_{IS} is the concentration of the cation-carrier complex in the membrane.

The partition parameter K depends on the equilibrium constant, K_{ex}, of the dominant extraction reaction Figure 25):

$$I^{z+}(aq) + zX^{-}(aq) + n\,S(m) = IS_n^{z+}(m) + zX^{-}(m) \tag{80}$$

and on the concentration of free ligands, c_s:[17,132]

$$K = (K_{ex} \cdot c_s^n)^{1/(z+1)} \tag{81}$$

It is obvious from Equation 79 that the concentration profile of the extracted ions in the membrane generally varies in space and time and, consequently, diffusion processes have to be considered at least in the subsurface membrane layers. Since the internal transport processes are rather slow compared to the rate of the transport processes in the aqueous boundary layer, the diffusion processes determine mainly the response time of carrier membranes.

Morf et al.[7] (Figure 26) suggested a suitable model which offers more insight into the dynamic behavior of neutral carrier membranes. According to this model, it was assumed that the delay of the transport due to the diffusion across the stagnant solution layer can be neglected in comparison to the processes in the membrane bulk. Thus, at time $t = 0$, whereby the induced diffusion of dissolved species through the membrane surface may start immediately:

$$\vec{j}\,'(t) = \vec{j}\,(0,t) \tag{82}$$

where $\vec{j}\,'(t)$ denotes the flux of free cations and anions within the boundary layer of the sample solution and $\vec{j}\,(0,t)$ represents the flux of cationic complexes and anions within the adjoining membrane boundary at $x = 0$, respectively. The flux within the outside diffusion layer may be approximated by the following relation:

$$\vec{j}\,'(t) = D' \frac{a_{i,2} - a'_i}{\delta} \tag{83}$$

Equations 82 and 83 show that the difference between the boundary a'_i and the bulk activity $a_{i,2}$ in the sample solution can be related directly to ionic diffusion within the membrane.

Membranes used in analytical practice can be treated with good approximation as infinitely thick diffusion layers during the first equilibration period. Accordingly, the flux into the membrane becomes[7,132,214,215]

$$j(0,t) = D \cdot \frac{c'_i - c_{i,2}}{\sqrt{\pi D t}} = D \cdot K \frac{a'_i - a_{i,2}}{\sqrt{\pi D t}} \tag{84}$$

where K is defined by Equation 79, and D is the mean diffusion coefficient of the electrolyte containing complexes $IS_N^{Z_i}$ and anions X^{Z_x} within the membrane assumed to be permselective (i.e., $D_{IS} \gg D_x$):[132]

$$D = \frac{(z_i - z_x)\, D_{IS} D_x}{z_i D_{IS} - z_x D_x} \approx \frac{z_i - z_x}{z_i} D_x \tag{85}$$

The dynamic behavior of carrier membranes may therefore be described with good approximation by the following relationship, obtained by combining Equations 82 to 84:

$$a_i' - a_{i,1} = (a_{i,2} - a_{i,1})\left(1 - \frac{1}{\sqrt{\frac{t}{\tau}} + 1}\right) \tag{86}$$

with the time constant:

$$\tau = \frac{DK^2\delta^2}{\pi D'^2} \tag{87}$$

Inserting Equation 86 into the Nernst equation, one obtains the time dependence of the potential function of neutral carrier electrodes following a step change in sample solution activity:

$$E(t) = E_2 + S \log\left[1 - \left(1 - \frac{a_{i,1}}{a_{i,2}}\right)\frac{1}{\sqrt{\frac{t}{\tau}} + 1}\right] \tag{88}$$

In agreement with the theoretical expectation, this potential vs. time function is basically different from the exponential time dependence found to be valid at high primary ion solution concentrations with membranes of constant membrane composition (see also Sections II.A, III.A and III.B), or from the hyperbolic response function used to describe the transient function of precipitate, based electrodes in dilute solutions (see Section III.C). This is clearly evident from comparison of Figures 7 and 27 and Tables 1 and 5.

It has to be pointed out, however, that for $t < \tau'$ (Equation 18), the rate of the mass transport in solution layer cannot be neglected, i.e., the short time behavior of carrier membranes is becoming similar to that of membranes with constant membrane composition (Equations 9, 10a, and 19). The simplified model leading to Equation 88 was extended later by Morf and Simon[132] assuming that the steady-state assumption at the membrane surface (Equation 83) is no longer retained, and the following potential time function for $t > \tau'$ was obtained:

$$E(t) = E_2 + S \log\left[1 - \left(1 - \frac{a_{i,1}}{a_{i,2}}\right)\left(\frac{1}{\sqrt{\frac{t}{\tau}}} + \frac{4}{\pi}e^{-t/\tau'}\right)\right] \tag{89}$$

whereas the general solution for $t < \tau'$ is

$$E(t) = E^0 + S \log\left[a_{i,1} + (a_{i,2} - a_{i,1}) \cdot \frac{2}{1 + \left(\frac{4\tau}{\pi\tau'}\right)^{1/2}} \operatorname{erfc} \frac{\pi}{4\left(\frac{t}{\tau'}\right)^{1/2}}\right] \tag{90}$$

Conversely, a square-root-type time dependence of the potential response is also found for membranes of constant composition if diffusion within the membrane cannot be neglected, e.g., in the presence of interfering ions.

The qualitative influence of the direction of the sample activity change on the response time is clearly the same for all types of membrane electrodes (Figures 7 and 27 and Tables 1 and 5). Accordingly, a considerably faster response can generally be expected when an

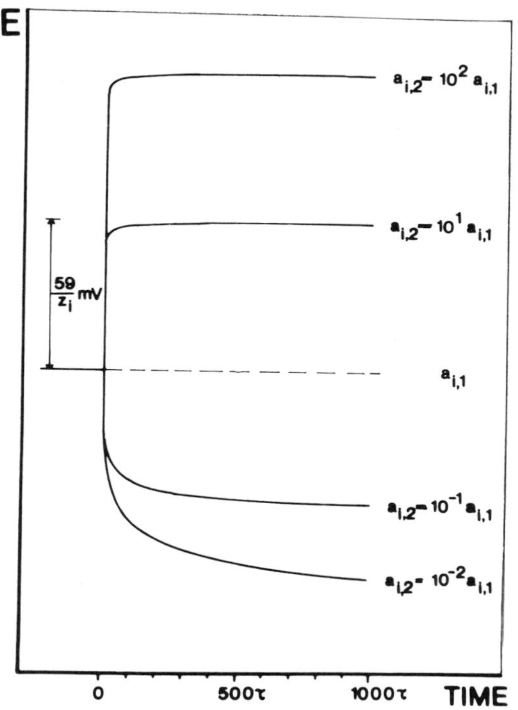

FIGURE 27. Theoretical EMF response vs. time profiles for neutral carrier membrane electrodes, calculated according to Equation 88.

Table 5
THEORETICAL RESPONSE TIME VALUES FOR CARRIER MEMBRANE ELECTRODES

Response time parameter	Values (sec) obtained from Equation 88 with τ = 1 msec	
	Activity increase ($a_{i,2}$ = 10 $a_{i,1}$)	Activity decrease ($a_{i,2}$ = 0.1 $a_{i,1}$)
50%	(0.0001)	(0.01)
95%	(0.05)	(5.29)
99.5%	6.03	6.03

Note: The values in parentheses may be unrepresentative: an extended diffusion model (Equations 89 and 90) led to higher values.[132]

activity increase is applied. A comparison of the theoretical time constants τ and τ' (Equations 18, 79, and 87) yields:

$$\tau \approx \tau' \frac{D}{D'} K^2 << \tau' \tag{91}$$

It follows from Equation 91 that, in contrast to membranes of constant composition, the

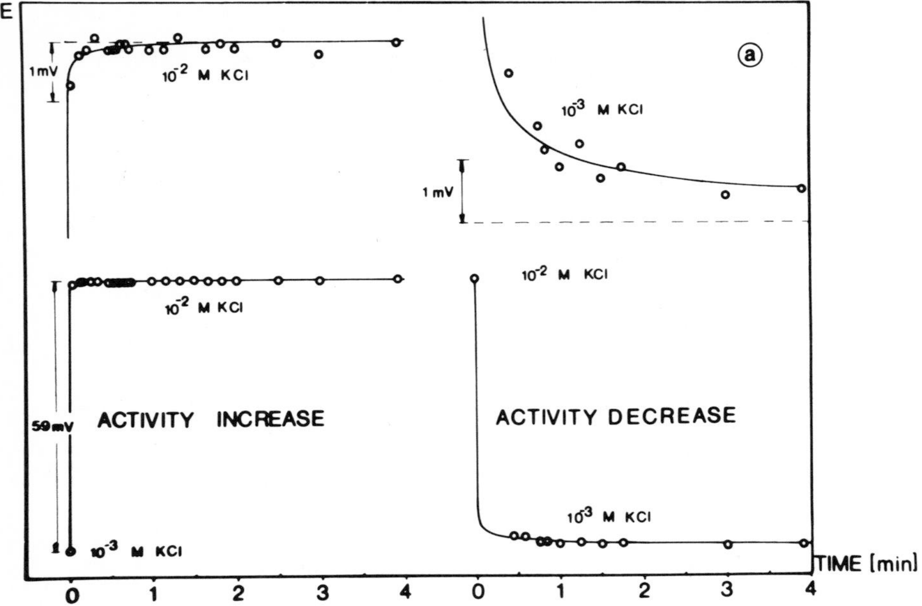

FIGURE 28. Influence of the activity change on the potential response vs. time profile of a K^+-selective carrier membrane electrode (nonpolar membrane; fast stirring). (a) Experimental points obtained from procedure A (large activity changes); theoretical curves according to Equation 88 with $\tau = 0.002$ sec. (b) Experimental points obtained from procedure B (small activity changes); theoretical curves according to Equation 88 with $\tau = 0.02$ sec.

dynamic response characteristics of carrier membranes are highly dependent on membrane properties such as extraction capacity (Equation 81) and resistance towards diffusion.

Summarizing the above treatment of carrier membrane electrodes, the following requirements must be met for a substantial reduction of τ, which is equivalent to a reduction of the response time:

1. Reduction of K (reduction of K_{ex} and c_s). The membrane components, i.e., solvent and matrix, should be as nonpolar as possible. The ion-selective ligand should be a relatively weak complex former, and its concentration in the membrane must be moderate. The sample solution should contain only lipid-insoluble anions.
2. Reduction of D. The mobility of sample anions in the membrane must be low. For a given membrane, the mobility of a species tends to decrease parallel to its lipid solubility.
3. Reduction of δ. The sample solution has to be stirred vigorously or a flow-through cell must be used. A reduced membrane surface is preferable.

A thin membrane is advantageous as far as response time is concerned.

To prove the validity of the above requirements based on the theoretical model, the effects of the direction of the activity change, of the properties of the membrane matrix, and those of the hydrodynamics of the system on the dynamic response characteristics of neutral carrier membranes were studied.

In order to study the influence of the direction of the sample activity change on the rate of response, the E vs. time profiles were measured at large and small activity changes (see Figure 28).[7] The experimental conditions were chosen to give the following changes in the steady-state cell potentials:[7]

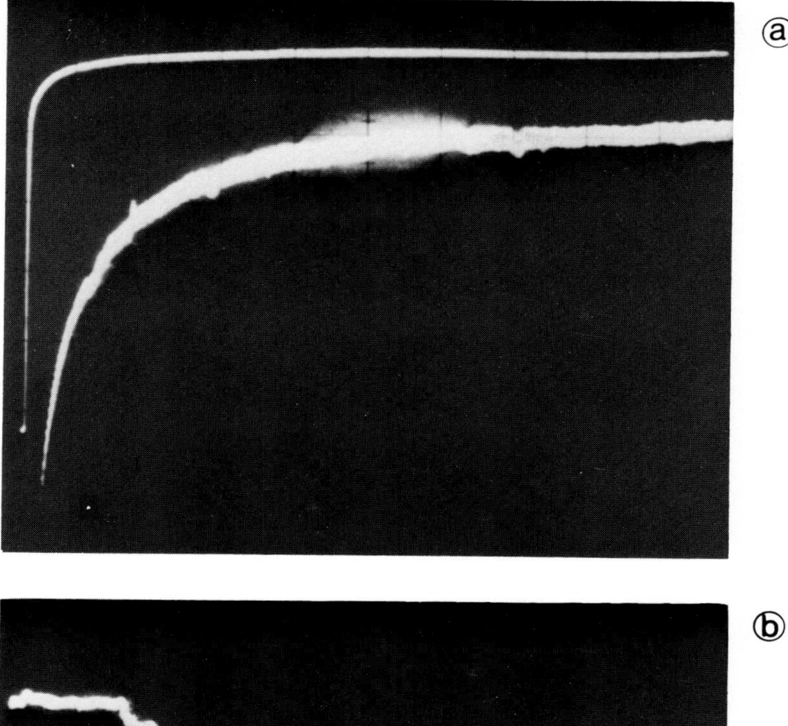

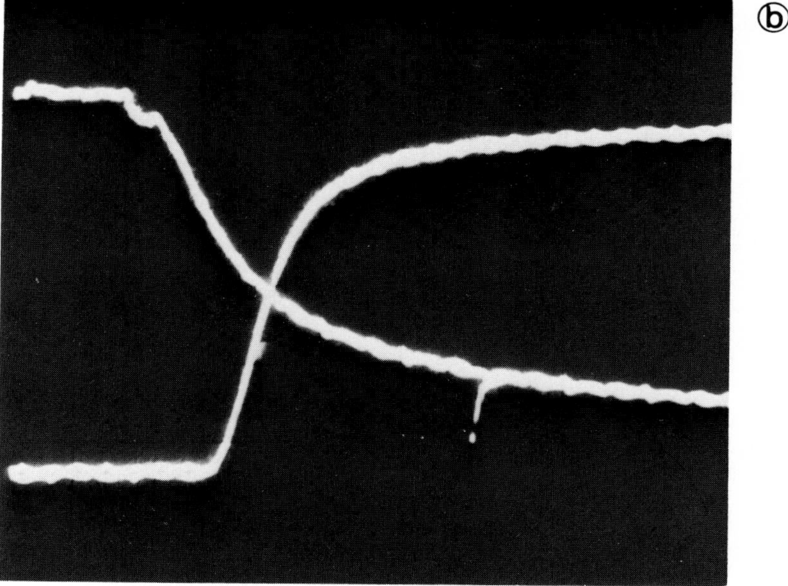

Plate 1. Response time curves of a valinomycin-based potassium-selective electrode marked K-2 in Table 7, and recorded in the experimental set-up shown in Chapter 2, Figure 16: $a_{i,1} = 10^{-3}$ *M* KCl; $a_{i,2} = 10^{-2}$ *M* KCl. (a) Upper curve: x = 1 sec/div, y = 10 mV/div; lower curve: x = 1 sec div, y = 1 mV/div (only the asymptotic part of the curve is shown in the picture). (b) Curves are recorded at increasing and decreasing concentration jumps: x = 10 msec/div, y = 10 mV/div.

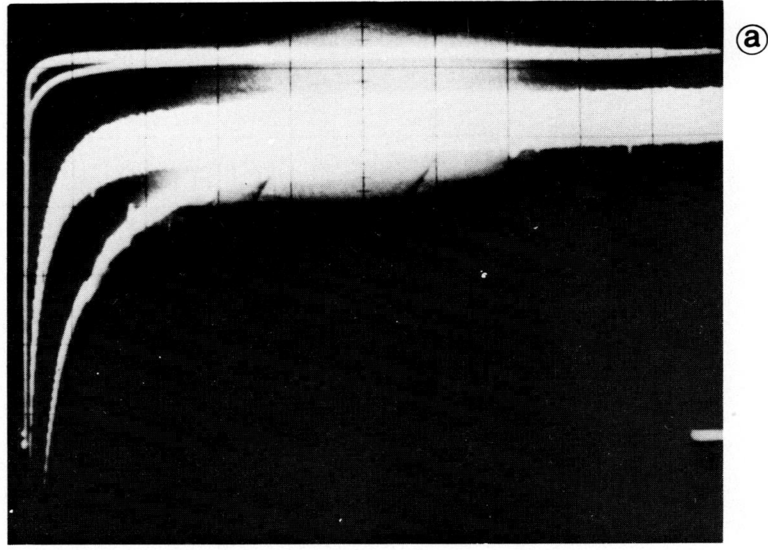

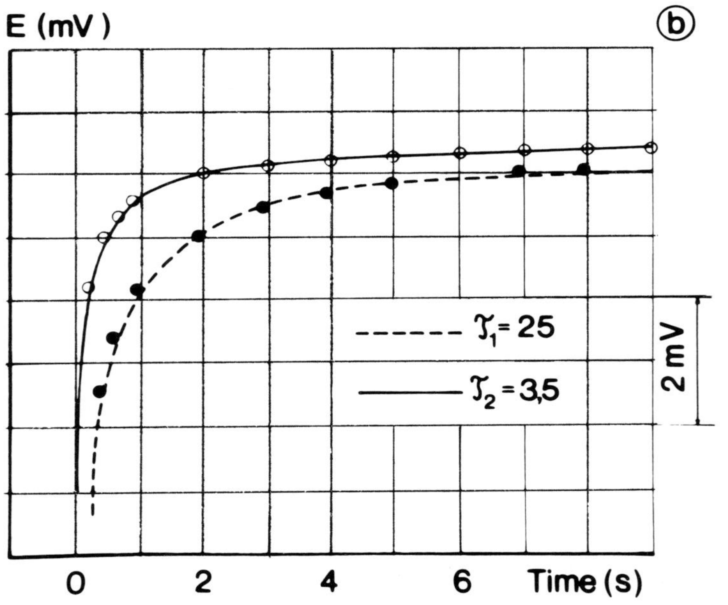

Plate 2. Response time curves of valinomycin-based potassium ion-selective membrane electrodes marked K-1 and K-2 (see Table 7) showing the "whole" and the "final" sections of the curves separately: $a_{i,1} = 10^{-3}$ *M* KCl, $a_{i,2} = 10^{-2}$ *M* KCl. (a) For the upper two curves: x = 1 sec/div, y = 10 mV/div. For the lower two curves: x = 1 sec/div, y = 1 mV/div. The first and third curves from the top correspond to electrode K-2, and the second and fourth ones to electrode K-1. (b) The curves are the same as the lower ones in (a) after suppression of noise. (——) Values measured with electrode marked K-2; (○○○) calculated with Equation 88 (τ = 3.5 msec); (— — —) values measured with electrode marked K-1; (●●●) calculated with Equation 88 (τ = 25 msec).

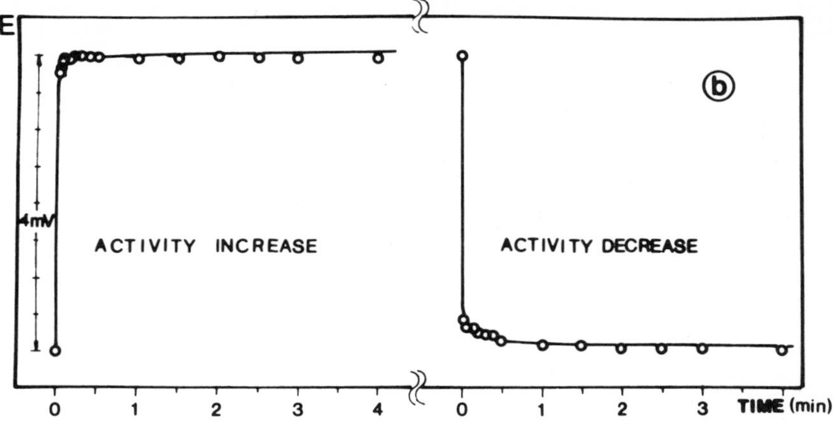

FIGURE 28b.

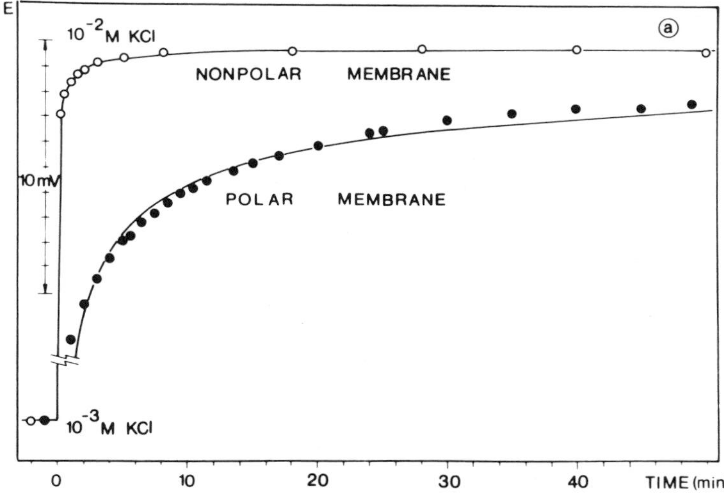

FIGURE 29. Influence of the membrane properties on the EMF response vs. time profile of a K^+-selective carrier membrane electrode. (a) Points from procedure A (activity increase; slow stirring); curves according to Equation 88 with $\tau = 0.4$ and 54 sec, respectively. (b) Points from procedure B (activity decrease; fast stirring); curves according to Equation 88 with $\tau = 0.02$ and 6 sec, respectively.

$$\text{Procedure A: } E_2 - E_1 = \pm 59 \text{ mV}$$

$$\text{Procedure B: } E_2 - E_1 = \pm 4 \text{ mV}$$

In excellent agreement with the theoretical predictions, the response was found to be considerably faster when the sample activities were changed from small to large values rather than the other way around.

The deviations of the time constants τ in Figure 28 were probably due to differences in the history of the membranes and in the measuring procedures applied.[7]

As predicted above [cf. (1) Figure 29; and Table 6], membranes with nonpolar solvents [DPP (dipentyl-phtalate): dielectric constant ~ 5] exhibited a much faster response than those with polar solvents [o-NPOE (o-nitrophenyl octylether): dielectric constant ~ 24].

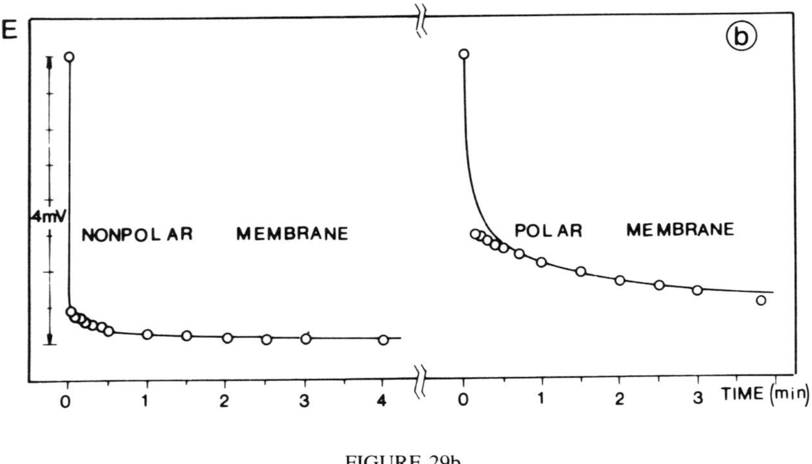

FIGURE 29b

Table 6
EXPERIMENTAL TIME CONSTANTS FOR DIFFERENT K^+-SELECTIVE ELECTRODES

Membrane composition[a] (%)			Matrix (residual percentage)	Time constants[b] (sec)		
Val	**NaTPB**	**Solvent**			τ	τ'
3.3	—	66.7 o-NPOE	PVC		6	—
3.3	—	66.7 DPP	PVC		0.02	—
2.2	—	70.2 DBP	PVC		0.02	—
2.4	0.66	70.5 DBP	PVC	KCl:	0.0035	—
				KSCN:	0.01	—
4.76	—	—	SR			0.225

[a] o-NPOE = 2-nitrophenyl octyl ether; DPP = dipentyl phtalate; DBP = dibutyl phtalate; PVC = polyvinyl chloride; SR = silicone rubber; Val = valinomycin; and NaTPB = sodium tetraphenylborate.

[b] All measurements were performed at high stirring[7] or high flow rates.[6,222]

Fast stirring of the sample solution reduces δ, and consequently the speed of response is increased [cf. (2)]. The experimental results shown in Figure 30 are in perfect agreement with this statement. Similar fast responses were observed at rather high sample flow rates.[222]

With silicone rubber membranes[6,222] at a rapid sample flow rate, both requirements (1) and (2) were met. Therefore, a considerable decrease in response time was expected. However, contrary to the expectations, the dynamic properties of silicone rubber-based electrodes are less favorable than the corresponding parameters of PVC electrodes (Figure 31; Table 6).

The normalized initial slope (at $t = 0$) of the response time curves of silicone rubber-based electrodes (cf. Chapter 5, Section III) was found to be only one fifth of that obtained for PVC membranes and one tenth of that recorded for PVC membranes containing tetraphenylborate as additive (see Figure 31; Table 7).

In addition, the transient functions of silicone rubber-based electrodes could not be described with Equation 88, derived for neutral carrier electrodes, but rather with Equation 19 (Figure 10), valid for membranes of constant composition.

The unusual dynamic character of silicone rubber membranes could be explained in terms

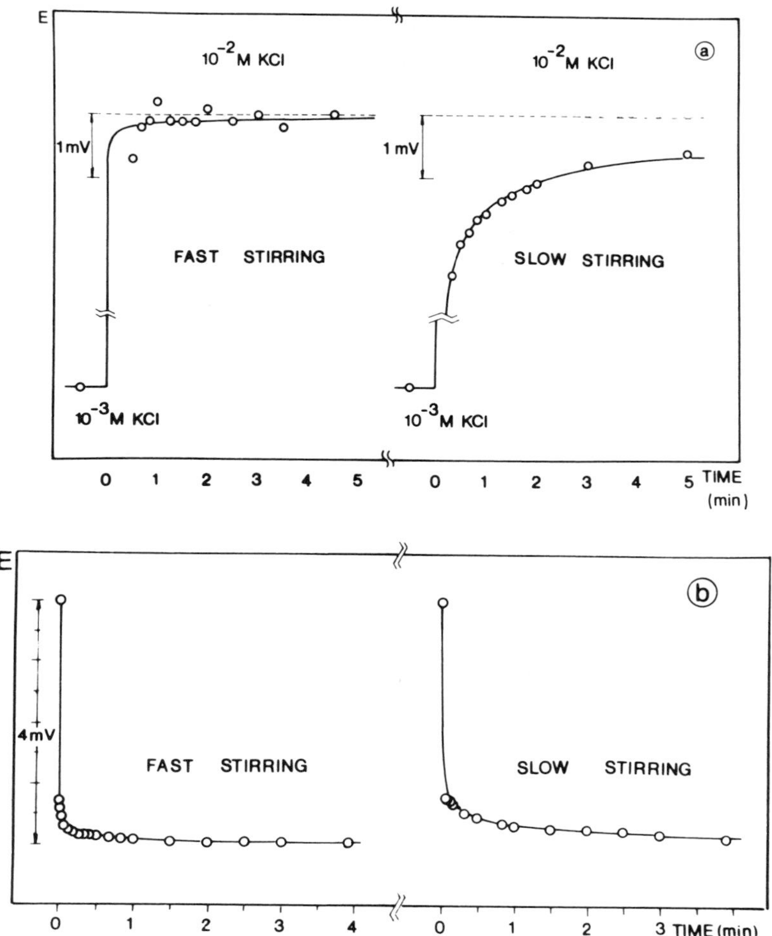

FIGURE 30. Influence of stirring on the EMF response vs. time profile of a K^+-selective carrier membrane electrode. (a) Points from procedure A (activity increase; nonpolar membrane); curves according to Equation 88 with τ = 0.002 and 0.4 sec, respectively. (b) Points from procedure B (activity decrease; nonpolar membrane; curves according to Equation 88 with τ = 0.02 and 0.45 sec, respectively.

of diffusion through a high resistance silicone rubber layer formed at the membrane surface (see Section II; Figure 8).

Since the introduction of the valinomycin-based K^+-selective electrode,[233,234] great efforts have been made to improve its characteristics.[235-237] The incorporation of ion-exchange sites (i.e., additives) into neutral carrier membranes [e.g., sodium tetraphenylborate (NaTPB)] has proved beneficial in many respects.[71,72,238-240] The additives reduced interferences due to lipophilic anions being in the sample,[71,72] significantly changed the selectivity,[17,71,132] boosted cation sensitivity in the case of carriers with poor extraction capability, and considerably lowered the electrical resistance of the membrane.[36-39,129] Some experimental curves are shown on Plates 1 and 2.* As shown in Table 7 and Plate 2, neutral carrier membranes containing tetraphenylborate exhibited more favorable dynamic characteristics than the unmodified neutral carrier-based membranes in agreement with theoretical expectations.[131,132,222]

* Plates will appear following page 72.

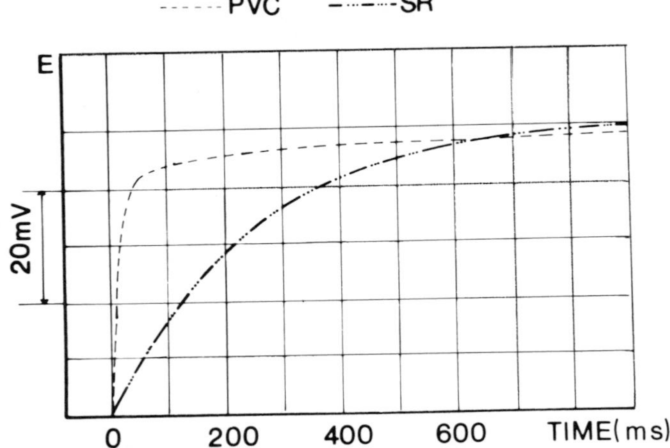

FIGURE 31. Response time curves of PVC (K-1)- and SR (K-SR)-based potassium ion-selective carrier electrodes (see Table 7): $a_{i,1} = 10^{-3}$ *M* KCl; $a_{i,2} = 10^{-2}$ *M* KCl.

Table 7
CORRECTED INITIAL SLOPES OF THE RESPONSE TIME CURVES OF POTASSIUM ION-SELECTIVE ELECTRODES

Matrix	Membrane composition[a] (%) Val	NaTPB	DBP	Code of the electrodes	No. of experiments	m_{eff} (mV/msec)
PVC	2.4	0.66	70.5	K-2	41	4.3 ± 0.3
PVC	2.2	—	70.2	K-1	42	1.9 ± 0.4
PVC	0.30	0.09	67.9	K-4	9	4.5 ± 0.6
PVC	0.28	—	67.8	K-3	9	2.0 ± 0.4
SR	4.76	—	—	K-SR	13	0.42 ± 0.04

Effect of the incorporated ion-exchange sites on the initial slopes of response time curves	$m_{eff\ (K-2)}m_{eff\ (K-4)} = 4.3/1.9 = 2.26$	$m_{eff\ (K-4)}m_{eff\ (K-3)} = 4.5/2.0 = 2.25$
Effect of the incorporated ligand concentration on the initial slopes of response time curves	$m_{eff\ (K-1)}/m_{eff\ (K-3)} = 0.97$	$m_{eff\ (K-2)}/m_{eff\ (K-4)} = 0.95$
Effect of the membrane matrix on the initial slopes of response time curves	$m_{eff\ (K-1)}/m_{eff\ (K-SR)} = 4.5$	$m_{(eff\ (K-2)}/m_{eff\ (K-SR)} = 1.02$

Note: Values were recorded in KCl solution.

[a] Val = valinomycin; NaTPB = sodium tetraphenilborate; and DBP = dibutyl phtalate.

The incorporation of NaTPB into the membrane resulted in an increase of the initial slopes of the transient function by a factor of 2 (Table 7), while the time constant of Equation 88 fitted to the data decreased to one fifth of its original value (Table 6).[222]

It is surprising that Equation 88 yields a fairly good fit to the response time curves of neutral carrier electrodes containing tetraphenylborate, although this electrode type can be assumed to preserve an approximately constant membrane composition.[131,132]

The favorable effect of the decrease in the concentration of the ionophore in the membrane could not be verified (see Table 7), contrary to theoretical expectations (cf. Equations 81, 87, and 88). However, in the presence of lipophilic anions, the response time was found to increase in agreement with theoretical considerations.[222]

V. UNIFIED MODELS FOR TRANSIENT FUNCTIONS

Several authors discussed critically the relative merits of various models interpreting the potential vs. time transient functions obtained by the activity step method.[17,132,134,151] The comparison is based on the following criteria:

1. The extent to which the model takes into account the effects of various parameters influencing the form of the transient functions
2. The extent to which the equations based on the theoretical models can be fitted to the experimental potential vs. time functions[226,241]

These studies revealed that the existing theories cannot describe the transient functions with satisfactory accuracy in the entire time domain.[134,136] This fact is due partly to the mathematical complexity of the problem and partly to the differences between the experimental conditions and the model assumptions. In fact, the explicit or implicit simplifications used in the derivation of the various models are only permissible in a certain time or concentration domain since the transient functions often result from parallel processes such as transport, adsorption, dissolution, crystallization, and charge transfer.[6] E.g., the hyperbolic relationship (Equation 72) of Buffle et al.[24,25] is only valid for dilute solutions and for $t \geqslant 1$ min. On the other hand, the model based on the assumption that diffusion through the hydrodynamic solution layer is the rate-determining process can only be accepted for relatively concentrated solutions and low flow or stirring rate (cf. Section II.A).[152]

Fleet et al.[139] (Equation 92), Shatkay[134] (Equations 15 and 93), and Belijustin et al.[242,243] suggested potential time functions containing several exponential terms with different time constants in order to obtain satisfactory fitting to the experimental response time curves in the entire time domain:

$$E(t) = E_1 + (E_2 - E_1)\,(1 - e^{-k_1 t} - e^{-k_2 t} - \cdots) \tag{92}$$

$$E(t) = E_1 + \sum_n \Delta E_n (1 - e^{-t/\tau_n}) \tag{93}$$

The empirical Equation 92 can advantageously be used for practical purposes, although it was not justified theoretically. However, Equation 93 can be meaningful in the linear regime (i.e., applying small activity steps; see Chapter 2, Section I.C).[10,22] An example of the application of Equation 93 is given in Figure 32.

Similarly, the summation of hyperbolic functions (e.g., Equation 72) also permitted a better fitting of the experimental curves:[134,226]

$$E(t) = E_1 + \sum_n \Delta E_n \frac{K_n t}{1 + K_n t} \tag{94}$$

where

$$E_2 - E_1 = \Delta E = \sum_n \Delta E_n \tag{95}$$

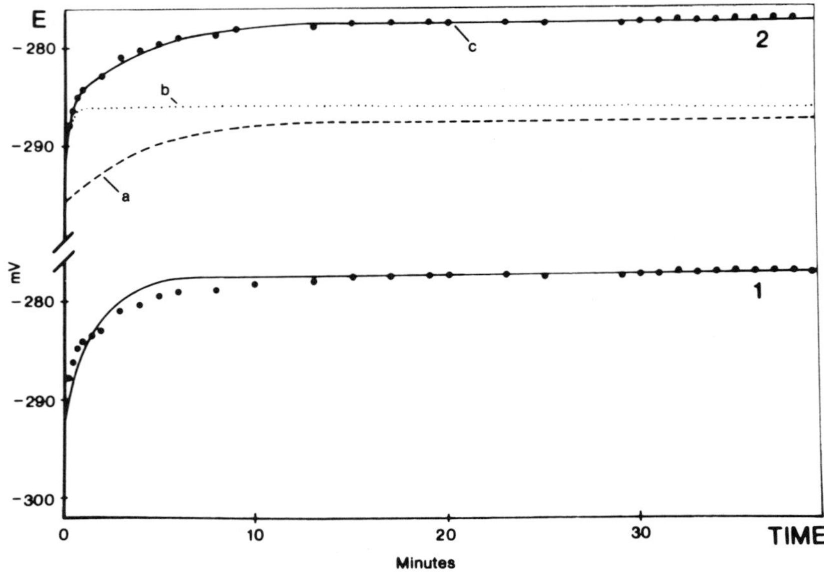

FIGURE 32. Potential response of a Ag_2S electrode from $AgNO_3$ $10^{-4} \rightarrow 10^{-5}$ *M*. Points are experimental; curves are theoretical E_1 (exper) = −296 mV. Curve 1: best fit from Equation 93 (see also Chapter 2, Equation 15) using n = 1, E_1 = −292 mV, ΔE_1 = 15 mV, τ_1^{-1} = 0.6 min^{-1}; best E_2 (calc) = −277 mV, ς = 1.1 mV (SD). Curve 2: best fit from Equation 93 using n = 2, E_1 = −296 mV, ΔE_1 = 10 mV, τ_1^{-1} = 5,0 min^{-1}; ΔE_2 = 9 mV, τ_2^{-1} = 0.24 min^{-1}; best E_2 (calc) = −277 mV, δ = 0.3 mV (SD). (From Shatkay, A., *Anal. Chem.*, 48, 1039, 1976. With permission.)

These expressions, however, can only be considered multiparameter curve fittings unless the partial processes of the overall electrode reactions are clarified. According to the authors' opinion, the mathematical formulation of the transient functions must be aimed for the elucidation and quantitative description of the effects of partial processes influencing the dynamic functions. The mathematical expressions would permit:

1. The calculation of the equilibrium potential following the activity step from the initial section of the transient functions. The latter requirement is especially important in the following cases: application of ISE in flowing solutions;[244-248] measurements performed in very dilute solutions;[134,136,152,153,156,157] and measurements carried out with the use of relatively sluggish electrodes, e.g., gas and enzyme electrodes.[7,17,216-221]
2. Reliable comparison of the response time data determined under different conditions (different measuring techniques and experimental parameters). This also would allow the elaboration of the generally acceptable method of the determination of the response time.[128,228]
3. Collection of experience on the transient functions for the development of electrodes and cells of good dynamic characteristics.[6]

Hawkings et al.[136] attempted to take into consideration the effect of various parallel and consecutive partial processes of the overall electrode reaction by the summation of the expressions derived for each partial process. They carried out experiments with the LaF_3-based fluoride ISE in very diluted solutions. Four distinct processes were identified: (1) reaction process; (2) ion diffusion; (3) dissolution of LaF_3; and (4) calibration drift.

The "apparent fluoride" concentration produced in the reaction process was described by an empirical equation which can be considered as a combined one (cf. Equations 65 and 70):

$$a'_i = a_{i,2} + (a_{i,1} - a_{i,2}) \frac{1 - e^{-\lambda t}}{\lambda t} \quad (96)$$

Time constant λ was found to depend on the concentration and the direction of the concentration change, and its value was determined in each experiment with multiparametric curve fittings on the basis of four data points selected in the time range from 10 to 60 min. Experiments carried out in the time range from $t = 0.5$ to $t > 10^{-4}$ min and in the fluoride concentration range from 3.7×10^{-8} to 7.9×10^{-4} *M* showed that good fitting was usually obtained at $t > 10$ min.

The discrepancy between observed and calculated time response (Equation 96) in the shorter time region was ascribed to a diffusion mechanism, and was taken into consideration by inserting Equation 10b into Equation 96:

$$a'_i = a_{i,2} + P\frac{1 - e^{-\lambda t}}{\lambda t} + Q\left[\frac{4}{\pi}\sum_{n=0}^{\infty}\frac{(-1)^n}{2n + 1} \cdot \exp\left(-\frac{D\pi^2}{4\delta^2}(2n + 1)^2 t\right)\right] \quad (97)$$

or

$$a'_i = a_{i,2} + P\frac{1 - e^{-\lambda t}}{\lambda t} + Q[1 - f(t)] \quad (98)$$

and at approaching $t \rightarrow 0$:

$$a_{i,1} = a_{i,2} + P + Q \quad (99)$$

Equation 97 provided a very good agreement between the experimental data and the calculated curve if the time was longer than 1 min.

The effect of dissolution on the apparent fluoride activity could be taken into consideration with the intorudction of a linear term, in Equation 97, while two additional exponential terms were introduced in order to account for the calibration drift:

$$a'_i = a_{i,2} + P\frac{1 - e^{-\lambda t}}{\lambda t} + Q[1 - f(t)] + kt + A_1 \cdot e^{-\alpha_1 t} + A_2 \cdot e^{-\alpha_2 t} \quad (100)$$

where α_1 and α_2 are time constants evaluated by curve fitting, and

$$a_{i,1} = a_{i,2} + P + Q + A_1 + A_2 \quad (101)$$

The relationship derived by Hawkings et al.[136]certainly shows an improvement compared to that of Shatkay[134] and Belijustin et al.,[242,243] as the former takes into account each partial process individually, while the latter considers only partial processes which can be expressed by exponential functions (cf. Equations 93 and Chapter 2, Equation 15).

Naturally, a sufficiently large number of exponential terms in Equation 93 permit the approximation of experimental curves with the required accuracy, although such a multiparametric curve fitting does not provide information about the physical phenomena. Unfortunately, neither relationship can interpret the physicochemical meaning of ΔE_n, P, Q, A_1, and A_2 in Equations 93 and 100. Nevertheless, the above relationships as well as similar multiparametric curve fittings may be very advantageous in the application of electrodes for the solution of analytical problems as shown by Hawkings et al.[136]

Summing up, it can be concluded that for describing the transient functions of ISE, a general model should be set up at first which permits the interpretation of various parameters, e.g., membrane properties, flow conditions, and primary ion activity. As a further step, a mathematical equation can be derived with appropriate boundary conditions based on this model. By a judicious selection of the experimental conditions, the assumed boundary conditions can be met, thus some partial processes of the overall electrode reaction can be studied individually (cf. Sections II.A and II.B).

Some possibilities of the complex formulation of the overall electrode process have been given in Section III.B dealing with consecutive reactions and in Section IV describing the ionophore-based electrodes (Equations 69, 89, and 90). A similar approach to the former was suggested by Hawkings et al.[136]

On the other hand, erroneous conclusions can easily be drawn when an appropriate physical model is sought to correspond to a mathematical relationship which is apparently suitable to describe the transient functions.

According to Shatkay and Hayano,[249] a general relationship of the transient functions of ISE must also account for the potential oscillations which appear on the response time curves in certain cases.[134,193,249] Thus, Shatkay and Hayano[249] suggested the following differential equation:

$$\frac{d^2E}{dt^2} + A\frac{dE}{dt} + BE = 0 \tag{102}$$

where A and B are positive constants.

In spite of the fact that a ''considerable number of physical models can be coupled to the mathematical model'' of Shatkay and Hayano, this model does not contribute to the fundamental understanding of transient potentials.[249]

Chapter 4

TRANSIENT POTENTIALS IN THE PRESENCE OF INTERFERING IONS

I. INTRODUCTION

In the previous chapters, ion-selective electrodes (ISE) have been subdivided into two groups on the basis of dynamic characteristics. The first group includes the electrodes with membranes of constant composition, while the second group contains the ionophore-based electrodes. With ionophore-based electrodes, the diffusion within the membrane bulk is the rate-determining step, while with electrodes belonging to the first group, it is difficult to find experimental conditions that permit the differentiation between the ion transport and other electrode processes because they have comparable rates (see Chapter 3, Sections II.A and II.B). Under extreme experimental conditions (high solution flow rates and highly diluted solutions), one may be able to distinguish between the partial processes of the overall electrode reaction.[152,153] Similarly, it may be advantageous if transient measurements are carried out in the presence of interfering ions in the so-called "two-ion range".[55,137,139,144,175,176,180,181]

In the presence of interfering ions, the transient functions are expected to be influenced by (1) the "competition" for the active sites of the electrode surface;[34] (2) the transformation of the electrode surface in time;[183,193,250] and (3) diffusion and the chemical reaction within the electrode membrane.[183,251-253]

The term "two-ion range" indicates that the equilibrium potential is determined by both the primary and interfering ion activities according to the Nicolsky equation:

$$E(a_i, a_j) = E_i^o + S \log (a_i + K_{i,j}^{pot} a_j^{z_i z_j}) \tag{1}$$

In the two-ionic range, the following inequality holds:

$$\frac{a_i}{K_{i,j}^{pot} a_j^{z_i/z_j}} \geqslant 1 \tag{2}$$

Thus, the difference in the equilibrium potential observed in the presence and the absence of interfering ions, respectively, is maximum 18 mV for $z_i = 1$:

$$18 \text{ mV} \geqslant \Delta E = S \log \frac{a_i + K_{i,j}^{pot} a_j^{z_i/z_j}}{a_i} \tag{3}$$

where $E(a_i, a_j)$ is the cell voltage measured in the presence of interfering ions; a_i and a_j are the activity of the primary and interfering ions, respectively; $K_{i,j}^{pot}$ is the potentiometric selectivity coefficient; z_i, z_j are the charge of the primary and interfering ions, respectively; and $\Delta E = E(a_i, a_j) - E(a_i)$ is the difference between the cell voltage measured in the presence of the interfering ion and that measured in the solution containing the primary ion only.

These types of studies give a better insight into the mechanisms of ion selectivity, in addition to providing data on electrode kinetics.[174-187] Besides this, useful practical analytical information can be obtained from response time measurements performed in the presence of interfering ions, since the ratio of the interfering ion and the primary ion can also vary during analytical determinations. The transient functions recorded in the two-ionic range are especially important in flow-through analytical techniques employing ISE.[149,248,254]

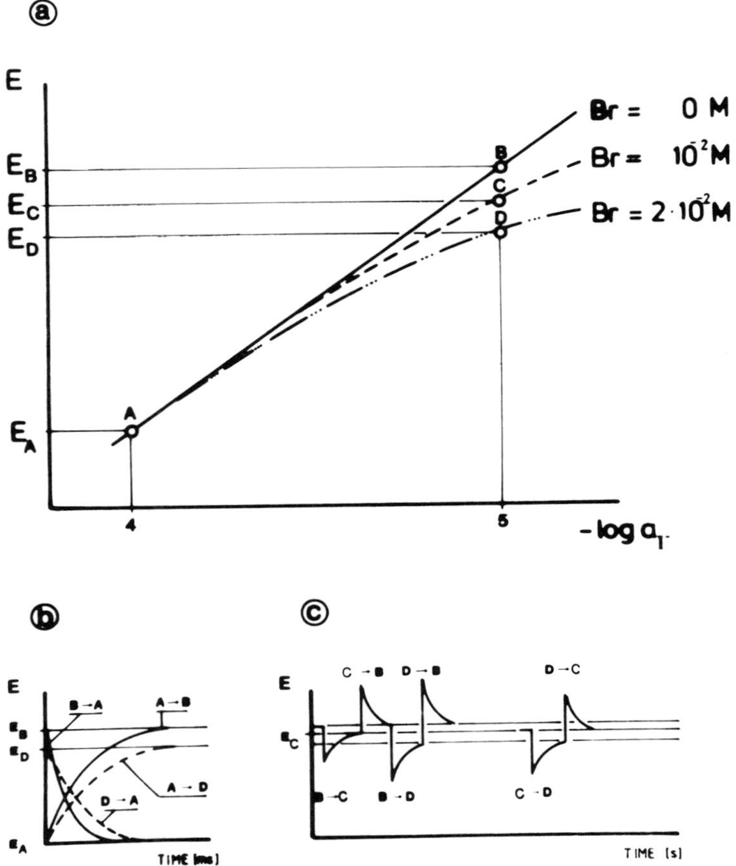

FIGURE 1. A hypothetical calibration graph at different activity levels, and the corresponding transient signals obtained at various activity steps. (a) Calibration graph of an iodide-selective electrode in the range of 10^{-4} to 10^{-5} *M* I^- in the presence of different interfering ion activities: $C_{Br} = 0$ (A-B), $C_{Br} = 10^{-2}$ *M* KBr (A-C), $C_{Br} = 2 \times 10^{-2}$ *M* KBr (A-D). (b) Transient signals recorded at primary ion activity step in the presence of constant interfering ion level. (c) Transient signals recorded at interfering ion activity step in the presence of constant primary ion level.

The possible activity steps in the two-ion range and the corresponding transient signals are shown schematically in Figure 1. Thus, response functions in the two-ion range can be recorded:

1. By introducing a primary ion activity step at constant interfering ion activity[135,139] (Figure 1: A $\rightleftharpoons$ C, A $\rightleftharpoons$ D)
2. By introducing an interfering ion activity step at a constant primary ion activity level (Figure 1: B $\rightleftharpoons$ C, B $\rightleftharpoons$ D, C $\rightleftharpoons$ D)
3. By altering both the primary and interfering ion activities[135,144,180]

The response time curves recorded in the two-ion range often show an unexpected, nonmonotonic character, e.g., fast potential overshoot followed by slow relaxation.

The recorded transient signals can be characterized with parameters such as ΔE_1, ΔE_2, ΔE_3, $t^1_{1/2}$, $t^2_{1/2}$ (Figure 2). ($\Delta E_1 - \Delta E_2$) or ($\Delta E_3 - \Delta E_2$) are characteristic for the overshoot

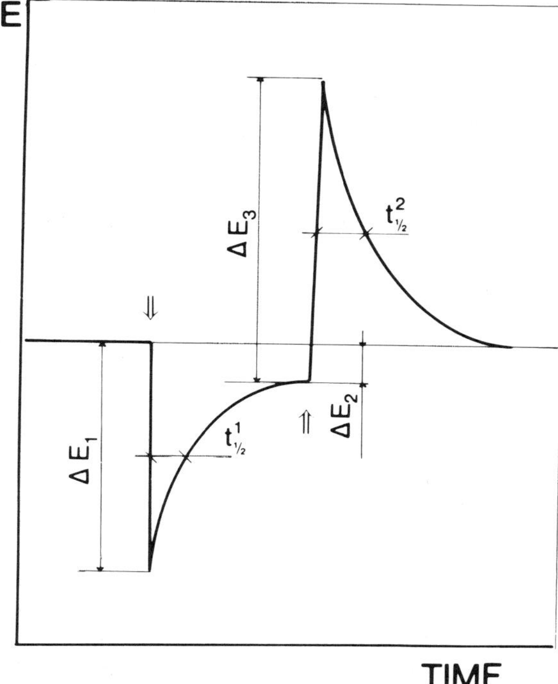

FIGURE 2. Parameters used to characterize the transient signals in the two-ion range.

observed following an increase or decrease of the interfering ion activity, respectively. ΔE_2 is related to the selectivity of the electrode,[255] while $t^1_{1/2}$ and $t^2_{1/2}$ are a measure of the rate of the transient response. Naturally, the value of $t_{1/2}$ is determined by the time constant of both the running up and the relaxation section.

Only a few papers have been published in the literature on these nonmonotonic transient signals. Those reports are related mainly to glass[144,177-179,187] or liquid ion-exchanger electrodes[55,175,176] and only recently to precipitate-based electrodes.[136,180]

Cammann[34] qualitatively explained the nonmonotonic transient signals with the exchange kinetic model (cf. Chapter 2, Section III.B). In this model, the exchange current density of the interfering ions is negligible compared to that of the primary ions. However, the "pre-equilibrium current density" of the interfering ions after the interfering ion activity step can often be many times larger than the exchange current density. Thus, a potential overshoot is observed in the preequilibrium transient region. According to Cammann,[34] the difference in the transient signals recorded in the presence of various interfering ions under the same experimental conditions is explained by the difference in the activation energies of the interfering ions. In agreement with the above statement, Karlberg[137] found that potential excursions arise when two ions compete to contribute to the potential, and thus one of the ions is inferior to the other in this process. Consequently, the ion exchange processes in the boundary regions were thought to be decisive. Later, Belijustin and others[144,242,243] explained that the transient signals in the two ion range resulted from the different mobilities of primary and interfering ions in the membrane phase. Rechnitz[179] suggested that the two effects combine to give the transient response, namely the relative ion-exchange equilibrium constant for the two ions and their mobility ratio.

II. THE SEGMENTED MEMBRANE MODEL

The transient response of ISE in the two-ionic range was interpreted mathematically first by Morf[174] on the basis of the segmented membrane model of Doremus.[256] Morf showed that inhomogeneities of the ion-selective membranes can cause sluggish or nonmonotonic transient behavior of the corresponding electrodes. The simplest way to treat inhomogeneities of the membrane phase is to consider the membrane a composite of homogeneous layers with different properties.[174,256] In fact, the study of the structure of hydrogen and alkali ion-selective glass electrodes in contact with aqueous solutions revealed that two layers of different properties, namely, a hydrated surface layer and a nonhydrated membrane bulk, can be distinguished.[137,201-203]

The potential of a cation-selective glass electrode in the presence of interfering ions is given by Morf[174] on the basis of the two-layer membrane model:

$$E = \text{const} + \frac{RT}{F}\ln\left[a'_i + K^{pot}_{i,j}(s)a'_j\right] + \frac{RT}{F}\ln\frac{c'_i + \dfrac{u_jK^{pot}_{i,j}(m)}{u_iK^{pot}_{i,j}(s)}c'_j}{c'_i + \dfrac{u_j}{u_i}c'_j} \tag{4}$$

where a'_i, a'_j = the activity of primary (I^+) and interfering (J^+) cations in the sample solution in contact with the membrane surface (at $x = 0$); $K^{pot}_{i,j}$ (s) = the potentiometric selectivity coefficient of the surface layer (thickness δ) of the membrane; $K^{pot}_{i,j}$ (m) = the potentiometric selectivity coefficient of the bulk of the membrane; c_i, c_j = the primary and interfering cation concentrations in the membrane surface layer at the internal boundary ($x = \delta$); u_i, u_j = the cation mobilities in the membrane surface layer; and R, T, F = their usual meanings.

If the membrane properties are constant throughout the cross-section of the whole membrane, i.e., if $K^{pot}_{i,j}$ (s) = $K^{pot}_{i,j}$ (m), the the membrane potential is independent of the concentration distribution in the membrane bulk and the second logarithmic term of Equation 4 is equal to zero. However, if the membrane properties are inhomogeneous, the potential varies with the concentration profiles.

For the description of the transient functions of ISE observed in the presence of the interfering ions, the following assumptions were made by Morf:[174]

1. Ionic diffusion through the membrane surface layer (thickness δ) is a slow process compared to the outside equilibration, but rapid compared to diffusion into the bulk of the membrane.
2. The total concentration (c) of cations is kept constant throughout the membrane surface.
3. The mobility (u) is the same for all cations in the membrane surface layer.

(The key assumption that ions within the membrane phase have the same mobility was criticized by Belijustin et al.[242])

It follows from Morf's first assumption that the derivation of the transient functions after an activity step in the sample solution (a'_i and a'_j) can be given by the mathematical description of the time dependence of c'_i and c'_j. For simplification, it is assumed:

$$a'_i = a_{i,1} \text{ for } t \leqslant 0$$

$$a'_i = a_{i,2} \text{ for } t \geqslant 0$$

The concentration change of the cations (c'_i and c'_j) in the bulk of the membrane as a function of time is formulated as follows:

$$c'_i = c_{i,2}f(t) + c_{i,1}\,[1 - f(t)] \tag{5}$$

where $c_{i,1}$ and $c_{i,2}$ are the steady-state concentrations before and after the activity step:

$$c_{i,2} = c\,\frac{a_{i,2}}{a_{i,2} + K^{pot}_{i,j}(s)\,a_{j,2}} \tag{6}$$

$$c_{i,1} = c\,\frac{a_{i,1}}{a_{i,1} + K^{pot}_{i,j}(s)\,a_{j,1}} \tag{7}$$

and c is the total concentration of the cations in the boundary layer of the membrane [cf. assumption (2)], while f(t) is defined by Equation 10b in Chapter 3:

$$f(t) = 1 - \frac{4}{\pi}\sum_{n=0}^{\infty}\frac{(-1)^n}{2n + 1}\exp\left[-(2n + 1)^2\,\frac{\pi^2 D}{4\delta^2}\,t\right] \tag{10b}$$

The cell voltage measured before the activity change in the sample solution containing both the primary and interfering ions is given by the following formula ($t \leqslant 0$):

$$E(t < 0) = \text{const} + \frac{RT}{F}\ln\,[a_{i,1} + K^{pot}_{i,j}(m)\,a_{j,1}] \tag{8}$$

This voltage obviously depends on the properties of the bulk of the membrane. At $t = 0$, i.e., at the instant of the activity step, the cell voltage is given by Equation 9:

$$E(t = 0) = \text{const} + \frac{RT}{F}\ln\,[a_{i,1} + K^{pot}_{i,j}(m)\,a_{j,1}] + \frac{RT}{F}\ln\frac{a_{i,2} + K^{pot}_{i,j}(s)\,a_{j,2}}{a_{i,1} + K^{pot}_{i,j}(s)\,a_{j,1}} \tag{9}$$

which depends on the properties of the bulk and the boundary layer of the membrane. The final equilibrium value ($t \rightarrow \infty$) is also determined by the properties of the bulk of the membrane:

$$E(t \rightarrow \infty) = \text{const} + \frac{RT}{F}\ln\,[a_{i,2} + K^{pot}_{i,j}(m)\,a_{j,2}] \tag{10}$$

For the intermediate time range ($t > 0$) Morf's general solution reads:

$$E(t > 0) = \text{const} + \frac{RT}{F}\ln\,[a_{i,1} + K^{pot}_{i,j}(m)\,a_{j,1}] + \frac{RT}{F}\ln\left[\frac{a_{i,2} + K^{pot}_{i,j}(m)\,a_{i,2}}{a_{i,1} + K^{pot}_{i,j}(m)\,a_{j,1}} + \frac{a_{i,2} + K^{pot}_{i,j}(s)\,a_{j,2}}{a_{i,1} + K^{pot}_{i,j}(s)\,a_{j,1}}\,(1 - f(t))\right] \tag{11}$$

The above model permits the interpretation of the slow potential response observed in the presence of interfering ions[139,179] at a primary ion activity step (Figure 1: A → D, D → A), and also the nonmonotonic potential responses recorded at constant primary ion activity as an effect of an interfering ion activity step (Figure 1: B ⇌ C, B ⇌ D, C ⇌ D).[149,180]

Thus, when only the primary ion activity is varied by the activity step (i.e., $a_{j,1} = a_{j,2}$), then:

1. A rapid potential response can be expected if $K^{pot}_{i,j}$ (m) and $K^{pot}_{i,j}$ (s) $\ll 1$, i.e., if both the surface layer and the bulk of the membrane are selective to cation I^+ (the membrane is practically assumed to be homogeneous).
2. Slow potential response is obtained if $K^{pot}_{i,j}$ (m) $\ll 1$, but $K^{pot}_{i,j}$ (s) $\gg 1$, i.e., if only the bulk of the membrane is ideally selective to cation I^+.
3. Nonmonotonic potential response is expected if $K^{pot}_{i,j}$ (m) $\gg 1$, while $K^{pot}_{i,j}$ (s) $\ll 1$, i.e., if the membrane surface in contact with the solution is selective to cation I^+, while the bulk of the membrane is selective to cation J^+.

Obviously, no potential response is obtained if both $K^{pot}_{i,j}$ (m) and $K^{pot}_{i,j}$ (s) $\gg 1$.

The above model is very plausible for glass electrodes as it is experimentally proved that the properties of the boundary layer differ from those of the bulk.[201-203] The model is also valid for liquid ion-exchanger membranes of moderate selectivity, since it can be assumed that the characteristics of the surface layer will become different from those of the bulk due to ion exchange during use in solution containing interfering ions.[139,175,176] However, due to the crystalline structure of precipitate-based electrodes, the existence of layers of different properties cannot be assumed, except if new phases are formed during use, e.g., in the case of LaF_3-based fluoride ISE in alkaline medium.[257,258] In spite of this, nonmonotonic potential response was also reported in the case of both freshly prepared precipitate-based electrodes or those which had a renewed surface.[180]

Later, Morf[183] modified the segmented membrane model[174] on the basis of Hulanicki's[209,210,259] equation relating to the apparent selectivity (nonselectivity) coefficient of ISE, and derived an equation describing the time dependence of the apparent coverage factor (s), which is essentially also a description of the time dependence of the apparent selectivity coefficient. For this, the following points were considered.

The response of an anion-selective electrode in the presence of an interfering ion if both ions have the same charge ($z = -1$) is described with the Nikolsky equation:

$$E = E^0_i - \frac{RT}{F} \ln(a'_i + K_{i,j}a'_j) \tag{12}$$

In practice, however, the potentiometric selectivity coefficients are defined in terms of bulk sample activities (see Equation 1):

$$E = E^0_i - \frac{RT}{F} \ln(a_i + K_{i,j}a_j) \tag{13}$$

Diffusion-controlled differences between the activities measured with the electrode (a'_i, a'_j) and those corresponding to the bulk of the solution (a_i, a_j) may cause characteristic variations in apparent selectivity coefficients. The rate of this alteration primarily depends on the interfering ion activity.

The theoretical selectivity factor $K_{i,j}$, in the case of precipitate-based ISE, is the equilibrium constant of the precipitate exchange (ion-exchange) reaction:

$$MeI + J^- = MeJ + I^- \tag{14}$$

where Me^+ represents a metal cation and I^- and J^- are the primary and interfering anions, respectively. The selectivity coefficient can be calculated from the solubility products of the corresponding precipitates:

$$K_{i,j} = \frac{L_{MeI}}{L_{MeJ}} \tag{15}$$

The authors, dealing with the apparent selectivity coefficient, used Equation 15 and the surface activities of the primary and interfering ions for their calculation. Besides this, the introduction of the surface coverage factor (Equation 16) was necessary for the fitting of Equation 15:

$$K_{i,j} = \frac{L_{MeI}}{L_{MeJ}} = \frac{a'_i[MeJ]}{a'_j[MeI]} = \frac{a'_i}{a'_j}\frac{s}{1 - s} \tag{16}$$

where L_{MeI} and L_{MeJ} are the solubility products of the corresponding precipitates of cation Me^+ and anions I^- and J^-, respectively; K_{ij} is the theoretical selectivity factor, i.e., the equilibrium constant of Equation 14; [MeI] and [MeJ] are the respective activities of the precipitates on the electrode surface; and s is the apparent coverage factor,[209] which represents the molar fraction of MeJ formed in the surface layer of MeI membrane by ion-exchange or dissolution/precipitation reactions.

The measured (apparent) selectivity factors do not agree with the calculated data (Equation 15) when the surface activities (a'_i and a'_j) differ from those of the bulk of the solution. Such differences are always encountered after an activity step. However, the deviation of the measured selectivity factor from the calculated one is large only when the equilibrium of Equation 14 is shifted towards the formation of the components on the right-hand side, i.e., when the precipitate formed by the interfering ion is less soluble than the electrode material ($K_{i,j} \gg 1$).[183,209,210,259]

The electrode surface is always acting as a receptor for the interfering ion if $K_{i,j}\, a_j > a_i$. According to Hulanicki and Lewenstam,[209,210] the fluxes at the boundary layer in stationary conditions can be given by the following equation:

$$J_i = \frac{D'_i}{\delta}(a_i - a'_i) = -J_j = \frac{D'_j}{\delta}(a_j - a'_j) \tag{17}$$

where $D'_i = \dfrac{D_i}{\gamma_i}$ and D_i and γ_i are the diffusion and the activity coefficients, respectively.

By expressing a'_i and a'_j from Equations 16 and 17 and substituting them into Equation 12, the following relationship is obtained:

$$E = E^0_i - \frac{RT}{F} \ln \frac{K_{i,j}\,[a_i + (D'_j/D'_i)\, a_j]}{K_{i,j}(1 - s) + (D'_j - D'_i)\, s} \tag{18}$$

From the above, it is obvious that the apparent selectivity coefficient defined by the bulk activities (a_i, a_j) of the solution depends on the coverage factor. Thus, it varies between the value defined at s = 0, namely

$$K^{pot}_{i,j} = \frac{D'_j}{D'_i} \approx 1 \tag{19}$$

and

$$s_{eq} = \frac{K_{i,j}a_j}{a_i + K_{i,j}a} \tag{20}$$

defined by

$$K_{i,j}^{pot} = K_{i,j} \tag{21}$$

The coverage factor varies in time depending on the ratio of the primary and interfering ion concentrations.

On the basis of experimental results with precipitate-based electrodes, Morf,[183] Jaenicke,[192,193] and Schwab[250] assumed that the rate-determining step of the reaction given by Equation 14 (e.g., the formation of a loose AgBr layer on the surface of a Ag/AgCl electrode in solutions containing bromide ions) is the diffusion of the anions to the electrode surface across the solution boundary layer. Thus, the flux of the interfering anions is given by

$$J_j = \frac{n_{tot}}{A} \cdot \frac{ds}{dt} \tag{22}$$

where A is the surface area of the electrode in contact with the solution containing interfering ions, and n_{tot} is the number of moles taking part in the surface exchange reaction.

By combining Equations 16, 17, and 22, the rate of the change of the coverage factor is expressed:

$$\frac{ds}{dt} = \frac{D'_jA}{n_{tot}\delta} \frac{s(a_i + K_{i,j}a_j) - K_{ij}a_j}{s(K_{i,j} - D'_j/D'_i) - K_{i,j}} \tag{23}$$

Morf[183] obtained the following expression for the variation of the coverage factor in time by integration of Equation 23 between the limits $t = 0$ and $t \rightarrow \infty$:

$$[K_{i,j} - (D'_j/D_i)]\, s - K_{i,j} \frac{a_i + (D'_j/D'_i)a_j}{a_i + K_{i,j}a_j} \cdot \ln\left(1 - \frac{s}{s_{eq}}\right) = (a_i + K_{i,j}a_j)\, C \cdot t \tag{24}$$

where

$$C = \frac{D'_jA}{n_{tot}\delta} \tag{25}$$

The variation of the electrode potential in time can be calculated if the change of the apparent coverage as function of time is known.

The response functions of an AgCl membrane electrode in bromide ion solutions of different concentrations were studied, among others, by Hulanicki and Lewenstam.[49,192,193,209,210,259] The experimental values reported by Hulanicki and Lewenstam and the fitted curves derived by Morf (with the use of Equations 24 and 18) are shown in Figure 3.

The theoretical data are in good agreement with the experimental results. This fact appears to prove the validity of the model assumption under the given experimental conditions. Namely:

1. Surface activities (MeI) and (MeJ) can be taken approximately equal to the surface coverage, s and $1 - s$, respectively (Equation 16).
2. The rate of the transformation of the surface layers is controlled by the diffusion across the hydrodynamic boundary layer.

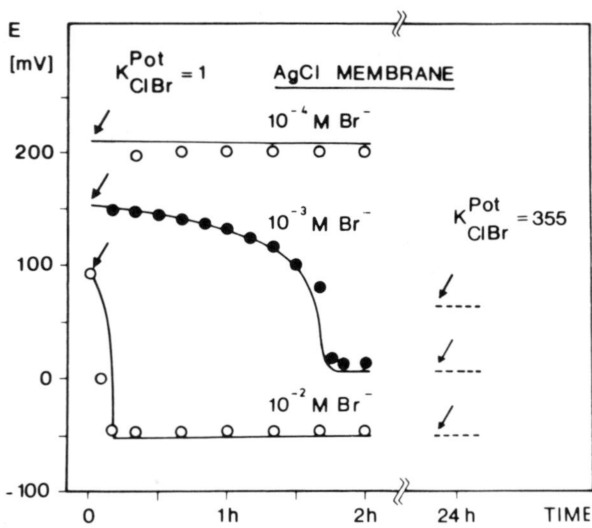

FIGURE 3. Time-dependent response of a silver chloride membrane electrode to bromide solutions of different concentrations. Solid lines: theoretical curves calculated from Equations 18 and 24 using $s_o = 0$, $D_{Br}/D_{Cl} = 1$, $\gamma_{Br}/\gamma_{Cl} = 1$, $K_{Cl,Br} = 355$,[72] and $C = 10\ \mathrm{min}^{-1}\ M^{-1}$. Points: experimental values (constant ionic strength) taken from Figure 1b in Reference 209. Dotted lines: theoretical curves calculated from Equations 13 and 26 using $\tau^* = 10^{-4}$ min M^2. (From Morf, W. E., *Anal. Chem.*, 55, 1165, 1983. With permission.)

Thus, taking as an example the silver chloride-based chloride ISE at the beginning of the contact with the solution containing interfering ions, the electrode indicates the activity of chloride ions generated from the membrane material in the precipitate ion-exchange reaction. Consequently, the apparent selectivity factor is almost equal to unity (Equation 19).

According to Morf's opinion, the mechanism of the dynamic response behavior shown in Figure 3 is very similar to that of the multilayer membrane model according to which differences in the selectivity coefficients of the bulk and the surface layer are responsible for the transient electrode behavior.[252] Thus, e.g., with silver chloride-based ISE, the selectivity factor for the membrane bulk is $K_{i,j}$ (Equation 15), while for the boundary layer (in the present case, it is assumed to be the hydrodynamic boundary layer) it is $D_j/D_i \approx 1$. Thus, the above model becomes formally suitable for the description of some phenomena observed with homogeneous, nonsegmented membranes in the two-ion range (slow or nonmonotonic potential response) (cf. Section III).

Basically different transient functions are expected in the two-ion range if the rate-determining step of the transformation of the ion-selective membrane is the diffusion inside the ion-selective membrane. This is especially characteristic of the so-called ion-exchanger-based liquid membranes.[17,183,252,253,258] Morf[17,252] derived an equation for the total ion activity measured by the electrode (cf. Equation 13) based on a liquid membrane approach of Jyo and Ishibashi:[253]

$$a_i + K_{i,j}^{pot}\, a_j = 1/2[a_i + K_{i,j}a_j - K_{i,j}\, f(t)] + \frac{1}{2}\sqrt{\left[a_i + K_{i,j}a_j - K_{i,j}\, f(t)\right]^2 + 4\, K_{i,j}\left(a_i + \frac{D_j'}{D_i'}\, a_j\right) f(t)} \qquad (26)$$

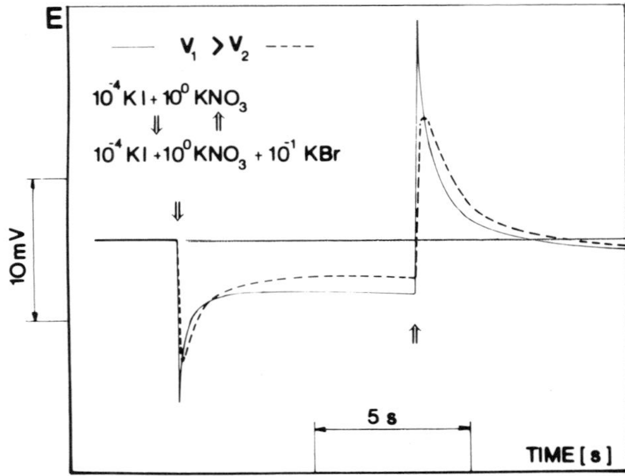

FIGURE 4. Effect of flow rate on the transient signal of an iodide ISE to a bromide ion activity step in the two-ion range.

where

$$f(t) = \frac{1}{\sqrt{t/\tau^*}} \quad \text{and} \quad \tau^* = \frac{D\delta^2 I^2}{\pi\, D'^2_j} \tag{27}$$

and D is the so-called average diffusion coefficient; I is the total anion concentration in the membrane; and δ is the thickness of the boundary layer.

A good agreement was found by Morf[17,252] between data calculated with the use of Equation 26 and experimental transient functions of liquid ion-exchanger calcium ISE at various calcium and magnesium ion concentration ratios.[139]

III. DIFFERENCES IN THE SURFACE ACTIVITY DUE TO ADSORPTION/ DESORPTION PROCESSES

Nonmonotonic dynamic response curves, similar to those reported for glass and liquid ion-exchanger-based electrodes, were recorded first for precipitate-based electrodes by Lindner et al.[180] with the AgI-based iodide electrode in the presence of bromide interfering ions. In the course of the study, the effect of flow rate (Figure 4) and the concentration of interfering ion (Figure 5) on the surface conditions of the electrode membrane (Figures 6 and 7) and the temperature of the sample solution (Table 1) on the parameters of the transient signals were studied. (Table 2).

A. Qualitative Interpretation

The interpretation of nonmonotonic transient functions on the basis of assumptions valid for systems consisting of layers of different selectivities (e.g., glass electrodes) does not seem to be appropriate for precipitate-based electrodes. The sensing membrane crystals have constant composition across the whole phase; the description of potential response of this type of electrodes does not require use of terms involving ''mobilities'' in the membrane phase. Consequently, a new model was elaborated on the basis of surface ''adsorption/ desorption'' phenomena. According to the model, as an effect of an interfering ion activity step, the primary ion activity is changed due to adsorption/desorption processes. As the electrode potential is determined by the surface activities, the transient potential functions reflect the activity change at the electrode surface in function of time.

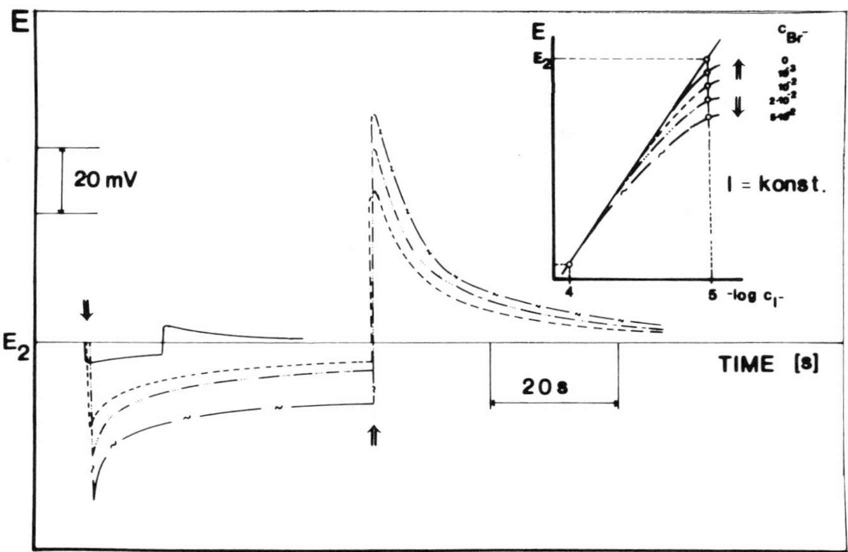

FIGURE 5. Effect of interfering ion activity on the transient signal of an iodide ISE due to a bromide ion activity step in the two-ion range.

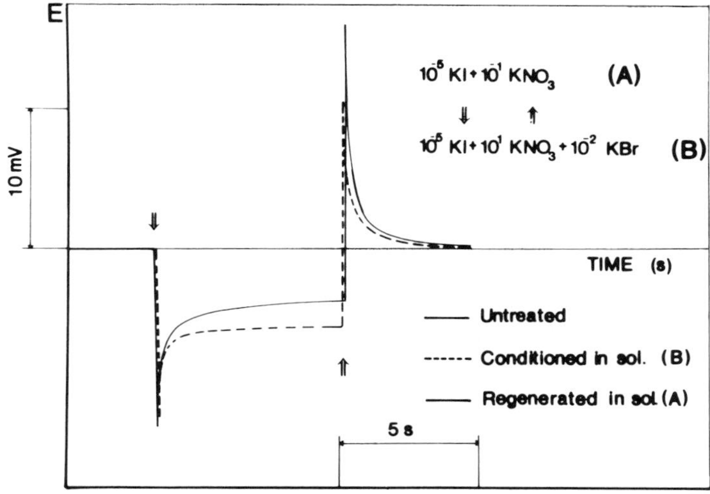

FIGURE 6. The effect of the pretreatment of the electrode surface on the transient signal observed in the two-ion range.

The differences in primary and interfering ion concentrations caused by the interfering ion activity step (Figure 1: B $\rightleftharpoons$ C, B $\rightleftharpoons$ D, C $\rightleftharpoons$ D) induce diffusion processes toward equalization. Accordingly, the transient signals are the result of the following processes taking place consecutively or simultaneously at the electrode solution interface (if the interfering ion activity is increased):

1. Diffusion of interfering ions (electrolyte) from the bulk of the solution to the electrode membrane surface
2. Adsorption (chemisorption) of interfering ions (being in large excess) on the electrode surface

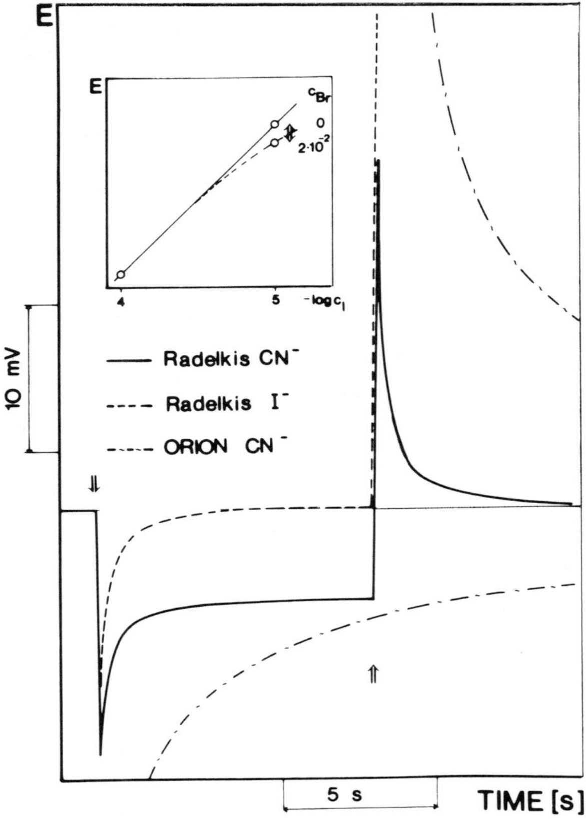

FIGURE 7. Transient signals in the two-ion range recorded with different types of ISE.

Table 1
THE MAGNITUDE OF THE POTENTIAL OVERSHOOT (ΔE_1) RESULTING FROM AN INTERFERING ION ACTIVITY STEP AS A FUNCTION OF TEMPERATURE AND STIRRING RATE

	ΔE_1	
	23°C	50°C
V_1	28.8 ± 3.5	37.8 ± 3.4
V_2	30.6 ± 4.2	40.4 ± 2.4

Note: $V_2 \approx 2V_1$. $\Delta C = 10^{-5}$ *M* KI + 1 *M* $KNO_3 \rightarrow 10^{-5}$ *M* KI + 1 *M* KNO_3 + 5 × 10^{-2} *M* K Br

Note: The experiments were carried out in a thermostated and magnetically stirred vessel containing 50 mℓ of a sample solution. The interfering ion activity jump was caused by injection of 1 mℓ of a solution containing 10^{-5} *M* KI + 2.5 *M* K Br.

Table 2
DESORPTION OF IODIDE IONS FROM AgI MEMBRANES IN DIFFERENT ELECTROLYTE SOLUTIONS

Solution composition used to promote desorption	Relative radiochemical activity of desorbed iodide ions[a]
0.1 *M* KNO_3	0.07
0.1 *M* KNO_3 + 10^{-3} *M* KBr	0.08
0.1 *M* KNO_3 + 10^{-2} *M* KBr	0.18
0.1 *M* KNO_3 + 5.10^{-2} *M* KBr	0.25
0.1 *M* KNO_3 + 10^{-2} *M* KBr + 10^{-5} *M* KI	0.39

[a] During the experiments, freshly prepared AgI pellets were soaked in 0.1 *M* KNO_3 containing 1×10^{-5} *M* KI labeled with ^{131}I isotope for 30 sec. The pellets were rinsed with distilled water, dried with filter paper, and the radioactivity of the pellets measured. Then the active pellets were placed in a solution for 1 min and the radioactivity of both the solution and the pellets measured (before the radioactive measurements, the treatment of the pellets was the same as before). The iodide amount desorbed and calculated from the increase of the radioactivity of the solution was practically the same as that calculated from the decrease of the radioactivity of the pellets. The relative radiochemical activity was defined as the ratio of radioactivity of solution and the pellet.

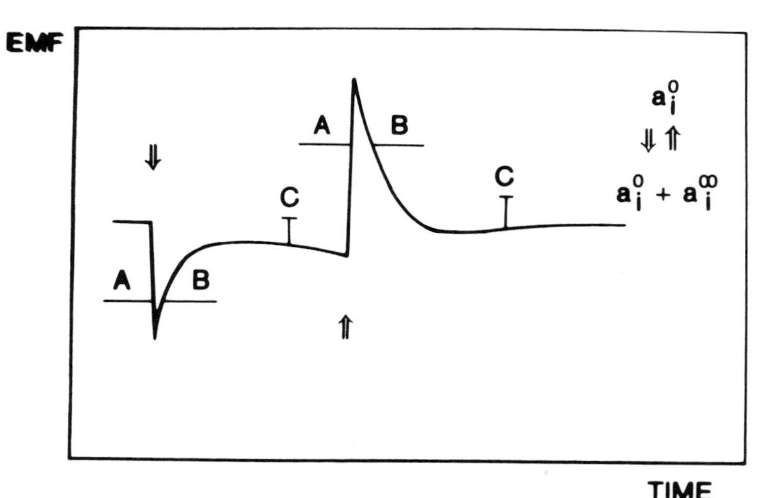

FIGURE 8. Typical nonmonotonic transient signals following interfering ion activity increase or decrease (a), and the corresponding potential determining processes at interfering ion activity increase (b). (A) Overshoot phase (0.1-sec range); (B) relaxation phase (second range); (C) slow surface transformation phase (minute range).

3. Desorption of primary ions from the electrode surface [parallel to process (2)]
4. Diffusion of excess primary ions into the bulk of the sample solution
5. Change of chemical composition and/or morphology of the electrode membrane surface and related diffusion processes

Processes (1) to (3) described above prevail in the fast increasing section of the transient signals (Figure 8: phase A; 0.1-sec range). However, during the decreasing relaxation section (Figure 8: phase B, range of seconds) and the slowest, drift-type signals range (marked C in Figure 8: range of minutes) process (4) or (5) prevails over the others.

Further assumptions are that (1) the sorption and ion-exchange reactions are much faster than transport processes, i.e., diffusion processess following the change in interfering ion

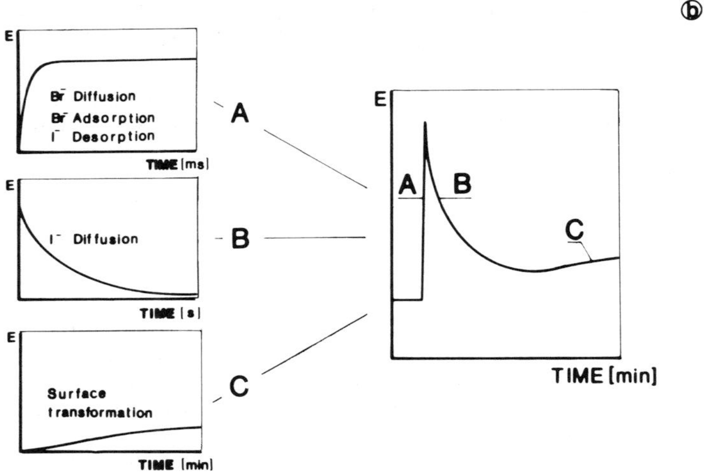

FIGURE 8B

activity; (2) the excess (or deficiency) of the primary ion levels off in a diffusional relaxation toward or from the bulk of the solution.

All experimental observations can easily be interpreted on the basis of the above model. It is not necessary to assume the existence of an outer membrane layer having a selectivity factor differing from that of the bulk of the membrane. Moreover, the model is found to be valid even when the mobilities of the primary and interfering ions are considered to be the same.

1. Effect of Flow Rate

All three sections of the transient signals considerably depend on the flow or mixing rate (Figure 4). Thus, it could be concluded (in contrast to some literature data[10,137,174]) that the relaxation of the transient signals occurs in the phase boundary solution layer.

Accordingly, the diffusion of the interfering ion is the rate-determining process during the fast increasing section of the transient signals, while in the relaxation period (section B, the potential decay), the diffusion of primary ions into the opposite direction controls the reaction rate. Both rates are greatly affected by the thickness of the stagnant diffusion film (cf. Chapter 3, Section II). Since both the second and third sections of the transient signals are flow rate dependent, ΔE_2 (used for calculating the selectivity coefficient[255]) also depends on the flow rate. The dependence of ΔE_2 on the flow rate indicates that the potential determining processes in the two-ion range cannot be considered equilibrium ones.[209,210,259] Thus, ΔE_2 can be used for the calculation of the theoretical selectivity data only with caution.[255]

2. Effect of Interfering Ion Activity

As shown in Figure 5, the magnitude of the transient signals (ΔE_1, ΔE_2, ΔE_3 in Figure 2) depends to a large extent on interfering ion activity. The experimental results indicate that the height of the overshoot is proportional to the logarithm of the concentration of the interfering ion with good approximation.

However, the signal height (peak height) is determined by the amount of primary ions generated in a given period of time in a given solution volume by the interfering ion. At any time instant, the amount of adsorbed and desorbed ions is defined by the actual activities of primary and interfering ions, in accordance with the appropriate adsorption isotherm.

3. Effect of the Direction of Activity Change

With the adsorption/desorption model, the effect of the direction of activity change (Figures 4 to 7) can also be explained; i.e., potential signals of the ISE are nonmonotonic not only at the appearance, but also at the disappearance of the interfering ion activity.

In accordance with the result of other workers,[144,175] the magnitude of the potential overshoot (ΔE_3) was found to be larger at decreasing interfering ion activity steps than at increasing ones (ΔE_1). Furthermore, $\Delta E_3 - \Delta E_2$ is also larger than ΔE_1 (where ΔE_2 is the steady-state potential difference between the potentials observed in the presence and the absence of interfering ions, respectively) (Figure 2). This can be explained as follows: at decreasing interference, the desorption of the interfering ions and the adsorption of primary ions take place at the electrode surface simultaneously with the related diffusion processes. The relative concentration decrease of primary ions in the solution boundary layer (of small volume) is much larger than the relative concentration increase in the opposite case, at otherwise identical absolute quantities of adsorbed or desorbed ions. As the electrode potential change is a function of relative concentration changes, it is evident that greater potential overshoot can be measured at decreasing activity steps using a logarithmic detector (ISE).

4. Effect of Electrode Surface Conditions

When the transient signals are studied in the two-ion range for a longer period of time, a third, very slowly changing section is observed which may partly be attributed to slow surface reactions. The standard potential of the electrode (E_i^0) and the selectivity factor (K_{ij}^{pot}) change continuously as a function of time in contact with the solution containing interfering ions (Figure 9).

In order to model the slow surface modifications of the electrode membrane, electrodes were pretreated in solutions containing both the primary and interfering ions, and were subsequently used for the corresponding dynamic experiments. The experimental results shown in Figures 6 and 9 clearly demonstrated that the pretreatment increased the value of ΔE_2 and the time constant of the transient signal, i.e., $t^1_{1/2}$ and $t^2_{1/2}$. The alteration of the electrode characteristics as a function of pretreatment time were found to be almost reversible under the experimental conditions studied.

The surface change of the electrode membranes was also modeled by depositing a small amount of silver bromide on the surface of a silver iodide membrane either by electrolytic methods or by pressing AgBr on the surface of an AgI pellet. After the measurement of the electrode response characteristics under stationary and dynamic conditions, the electrode surface was regenerated by conditioning the electrodes in pure potassium iodide solutions. The main parameters of the transient signals were measured as a function of the regeneration time.

The model experiments with silver bromide-covered iodide electrodes showed that the presence of a macroscopic amount of silver bromide on the electrode surface changed fundamentally both the equilibrium and the dynamic characteristics of the sensor. However, the dynamic characteristics of the sensor gradually approached that of the untreated silver iodide-based electrodes (Figure 10) with the progress of the dissolution of the silver bromide from the electrode surface during the regeneration experiments.

The effect of the quality and structure of the ion-selective membrane surface on the time constant and magnitude of the transient signals is shown in Figure 7 for various commercially available silver iodide-based cyanide electrodes. The transient signals of cyanide ion-selective membranes made of pure silver iodide or silver sulfide and silver iodide mixtures, which have been used for a longer period of time, differ to a much greater extent than is expected on the basis of equilibrium data.

The large time constant of the transient signal observed with an Orion cyanide electrode, used for a longer period of time, may be explained by the change of the surface structure

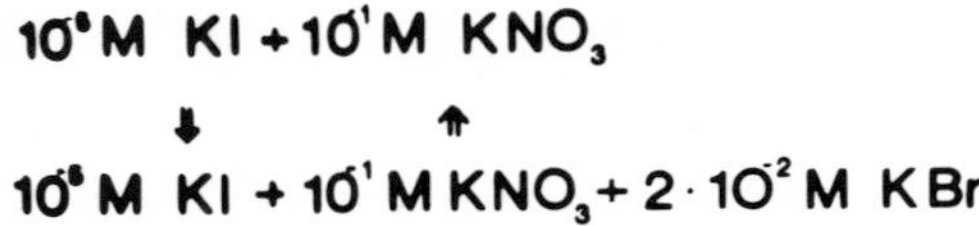

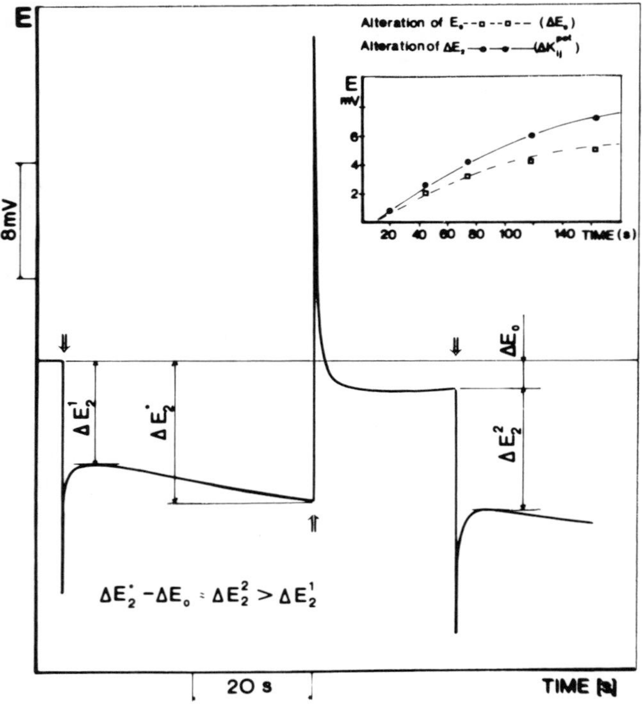

FIGURE 9. Nonmonotonic transient signals and the trend of the alteration of the standard potential (E^0) and ΔE_2 in function of time. The alteration of parameters E^0 and ΔE_2 was calculated from the dynamic response curves. ΔE^0 is the alteration of the standard potential E^0, i.e., the change of potential measured in pure primary ion solutions before and after the interfering ion activity step; $\Delta K_{i,j}^{pot}$ is the change of selectivity coefficient calculated from the change of the difference of potentials (ΔE_2) measured in the presence and absence of the interfering ions at a given primary ion activity level.

and composition of the electrode membrane material during use; i.e., in cyanide solutions, silver iodide is dissolved and a silver sulfide skeleton may remain, partially. Thus, the diffusion processes are hindered compared to those in the adhering solution layer at the use of pure silver iodide membranes.

5. Effect of Temperature on the Transient Signal

In contrast with the findings of Bagg and Vinen,[175] the processes determining the ΔE_1 values were found to depend on temperature (Table 1). The effect of temperature on the transient signals can simply be explained by the temperature dependence of sorption and diffusion processes.

The model discussed so far can be extended to interpret the nonmonotonic transient signals of other types of ISE,[180,187] i.e., glass and other ion-exchanger-based electrodes. However, if the interfering ion diffusion as well as the reversed primary ion diffusion take place within a hydrated gel layer, the rate of the processes which determines the transient signal is expected to decrease.[137] This can be attributed partly to the increase of the diffusion layer

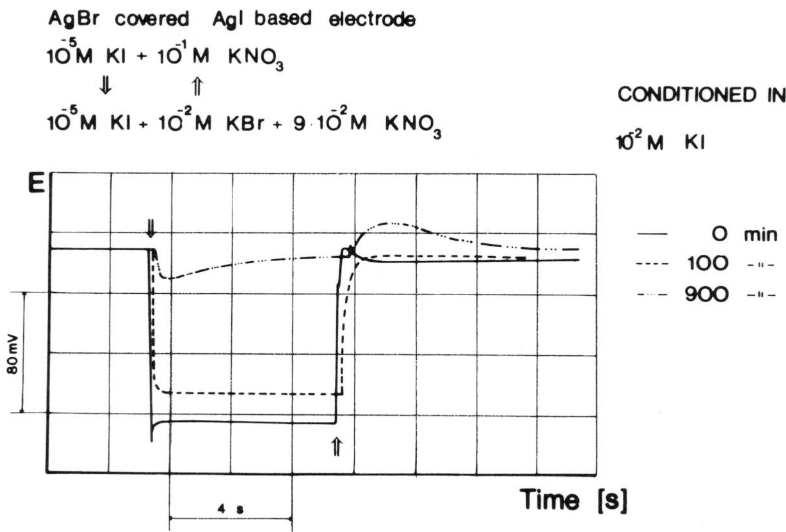

FIGURE 10. Transient signals recorded with silver bromide-covered silver iodide-based electrodes. For a better comparison, the changes in the potential values were plotted. The potential values of the freshly prepared and the regenerated electrodes (for 900 min in 10^{-2} *M* KI) in 10^{-5} *M* KI + 10^{-1} *M* KNO_3 were −100 mV, while that of an AgBr-covered iodide electrode was +30 mV.

thickness and partly to the decrease of the diffusion coefficient compared to that observed in the solution phase.

B. Quantitative Interpretation

The quantitative treatment discussed here is limited to the first two sections of the transient signals (Figures 8: phases A and B), since chemical changes of the membrane surface are rendered to the slowly changing section of the response curves (Figure 8: phase C). This treatment is not applicable when the cations or anions of the membrane material and the interfering ions form precipitates of lower solubility than that of the membrane material itself, because:

1. Nonmonotonic potential overshoot-type transient functions discussed above cannot be observed, but rather the opposite phenomenon occurs, i.e., the electrode potential varies according to a monotonic, asymptotic function (e.g., in the case of AgCl-based electrodes in the presence of Br^-or I^- ions).[49,183,192,193,209,210]
2. The membrane material changes irreversibly, and thus the membrane behavior becomes to a certain extent analogous to that determining the third phase of transient signals.[193,250]

The overall process is described by this model as follows (when a stepwise increase in interfering ion activity is considered):

1. At the time t = 0 and x = δ, the activity of the interfering ions is suddenly changed from $a_{j,1}$, to $a_{j,2}$ (Figure 11), and at the same time diffusion starts toward the electrode surface (x = 0), where the interfering ions are partly drained, due to adsorption processes, at a rate which varies with time.
2. An amount of primary ions, equivalent to that of the chemisorbed interfering ions, is desorbed parallel to the adsorption (chemisorption) of interfering ions present in large excess on the electrode surface. The quantity of adsorbed and desorbed ions corresponds to the appropriate adsorption isotherm in the two-ion range at the actual activities of primary and interfering ions in any time instant.

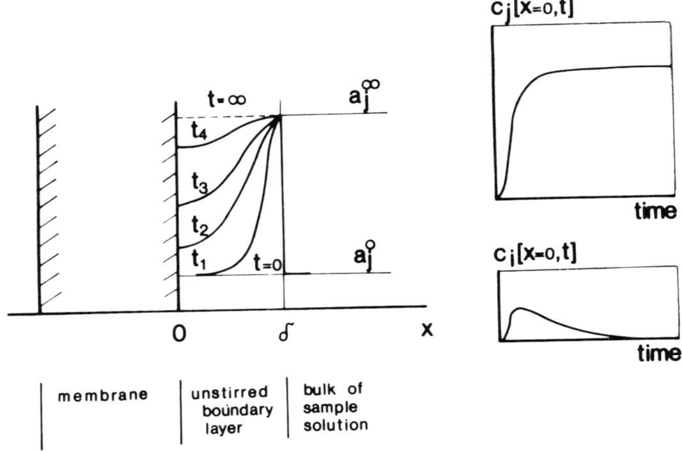

FIGURE 11. Schematic diagrams displaying the assumptions used for the mathematical descriptions: x, length coordinate perpendicular to the membrane surface (x = 0 at the surface, x = δ at the outer boundary of the diffusion layer); t, time (t = 0 at the stepwise change in the interfering ion activity). For other symbols, see the text.

3. The primary ions have a time-dependent source due to the adsorption of interfering ions. The bulk of the solution acts as a drain at x = δ for the primary ions desorbed from the electrode surface. The ion transport between the membrane surface (x = 0) and the bulk of the solution (x = δ) is controlled by diffusion.

The direction of diffusion and of adsorption and desorption processes changes sign when the activity of the interfering ions decreases, otherwise the transient functions can be described similarly as above.

In order to derive the nonmonotonic transient signals, the surface activities [a'_i (x = 0,t); a'_j (x = 0,t)] obtained by solving the differential equations describing the above-mentioned processes should be inserted into the Nicolsky equation (Equation 12). This is made difficult by the fact that the adsorption (chemisorption) isotherm, valid in the presence of both primary and interfering ions, is not known. Moreover, the boundary conditions of one of the differential equations (describing the transport of both primary and interfering ions) depend on the solution of that relative to the other ion.

As a first and simplest approximation, the desorption of the primary ions following the stepwise increase in interfering ion activity is assumed to occur instantaneously (although the diffusion of the interfering ions toward the electrode surface is not an instantaneous process). This assumption is supported by the fact that the concentration gradient of the interfering ions is several orders of magnitude larger than that of the primary ions desorbed and diffusing in the opposite direction. The boundary conditions of the differential equations used for describing the simultaneous counterflowing transport processes also differ fundamentally.

At this approximation, the amount of interfering ions chemisorbed on the electrode surface is neglected in the calculation of the surface activities because of the large excess of the interfering ions. This is justified by the fact that, in the Nicolsky equation (Equation 12), the interfering ion activity is multiplied by the weighing factor $K_{i,j} \ll 1$, and the relative (and not the absolute) activity changes of the interfering ion are controlling the nonmonotonic transient electrode response. In this case, the mathematical formulation of the problem is as follows (concentrations are used in the equations because the ionic strength is maintained constant in the experiments):

$$c_i(x,t = 0) = M_i\Delta(x) \tag{28}$$

$$c_i(x = \delta,t) = 0 \tag{29}$$

$$\frac{\partial c_i(x = 0)}{\partial x} = 0 \tag{30}$$

$$\frac{\partial c_i(x,t)}{\partial t} = D' \frac{\partial^2 c_i(x,t)}{\partial x^2} \tag{31}$$

$$E(t) = E_i^o - S \log [c_i^o + c_i(x = 0,t) + K_{i,j}c_j(x = 0,t = \infty)] \tag{32}$$

where x is the distance from the electrode surface; at the electrode surface, $x = 0$, and at the outer boundary of the diffusion layer, $x = \delta$; c_i^o is the bulk concentration of the primary ions; $c_i(x = 0,t)$ is the increase in the primary ion concentration referred to c_i^o in the vicinity of the electrode surface, due to desorption; c_j $(x = 0, t = \infty)$ is the final concentration of the interfering ions following the activity change; $K_{i,j}$ is the selectivity coefficient of the ISE with respect to the interfering ion j measured potentiometrically; E(t) is the cell potential measured as a function of time; S is the experimentally determined slope of the logarithmic calibration curve of the electrode; M_i is the quantity of primary ions forced to desorb from the electrode surface ($mmol/cm^2$); $\Delta(x)$ is the Dirac-delta function; and D' is the mean diffusion coefficient of ions in the adhering solution layer.

The solution of the differential Equation 31 with the boundary and initial conditions (Equations 28 to 30) with fast convergence in the short time range $\sqrt{D't} \ll \delta$ is as follows:

$$c_i(x,t) = \frac{M_i}{\sqrt{\pi D't}} \left\{ \exp\left(-\frac{x^2}{4D't}\right) + \sum_{k=1}^{\infty}(-1)^k \left[\exp\left(-\frac{(2k\delta + x)^2}{4D't}\right) + \exp\left(-\frac{(2k\delta - x)^2}{4D't}\right)\right]\right\} \tag{33}$$

Function 33 was derived in an elementary way with the help of the principle of reflection and superposition.[214,215] This function was found to converge in the entire time range when a sufficient number of terms was considered.

However, in the time range $\sqrt{D't} \ll \delta$, it is sufficient to use only the first term. The potential time function obtained by inserting Equation 33 into Equation 32 was fitted to experimental transient curves recorded in the two-ion range.

Some of the parameters (E^o,S) were determined by previous calibration; c_i^o and c_j $(x = 0, t = \infty)$ were given by the experimental conditions, while D' was taken to be a mean value of the diffusion coefficients of the two ions discussed: 1.86×10^{-5} cm^2/sec.[260] The $K_{i,j}$ M_i, and δ values were varied in order to obtain the best fit. $K_{i,j}$ was determined first from the steady-state potential values observed before and after the interfering ion activity step, and then the optimum values of M_i and δ were found on the basis of the peak height of the overshoot signals and of the relaxation rate of the curves, respectively. The results (e.g., Figure 12) justified that the simplifications in the mathematical derivation mentioned above were permissible. As far as the reliability of the fitted parameters is concerned, the

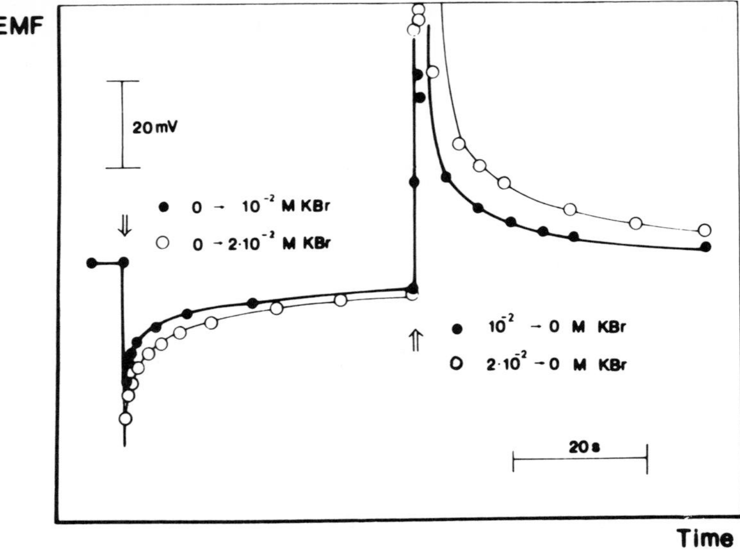

FIGURE 12. Time-dependent response of a silver iodide precipitate-based electrode in solutions containing 10^{-5} *M* KI and different bromide concentrations. Solid lines: theoretical curves calculated from Equations 32 and 33 using the following fitting parameters: $\delta = 2.9 \times 10^{-2}$ cm, $K_{i,j} = 2 \times 10^{-4}$, $M_i = 9.3 \times 10^{-5}$ mmol/cm^2 (for a bromide activity step; $0 \leftrightarrow 10^{-2}$ *M* KBr) and $\delta = 4.95 \times 10^{-2}$ (4.1×10^{-2}) cm, $K_{i,j} = 4.10 \times 10^{-4}$, $M_i = 1.8 \times 10^{-7}$ (1.4×10^{-7}) mmol/cm^2 (for a bromide activity step; $0 \rightarrow 2 \times 10^{-2} \rightarrow 0$ *M* KBr, respectively). Points: experimental values at constant ionic strength (10^{-1} *M* KNO_3). ●: activity step $0 \leftrightarrow 10^{-2}$ *M* KBr. ○: activity step $0 \leftrightarrow 2 \times 10^{-2}$ *M* KBr.

experimentally found sensitivities of the transient signals to $K_{i,j}$ and δ are reflected properly by the mathematical model. The effect of changes in M_i on the overshoot depends on the background primary ion level.

It was deemed important to check whether or not the assumption of instantaneous desorption of primary ions implies strongly disparate diffusion rates for the two ionic species.[183-185] In order to answer this question, it is sufficient to determine the quantity of primary ions desorbed from the electrode surface as a function of the time-dependent activity of the interfering ions. Thus, simultaneously, two diffusion processes (of finite rate and of opposite direction) are considered. In this way, diffusion rates are explicitly taken into consideration for both the primary and interfering ions. The corresponding mathematical derivation is the following:

$$c_i(x,t = 0) = 0 \tag{34}$$

$$c_i(x = \delta,t) = 0 \tag{35}$$

$$D' \frac{\partial c_i(x,t)}{\partial x} = -\frac{\partial M_i}{\partial t} \tag{36}$$

$$\frac{\partial c_i(x,t)}{\partial t} = D' \frac{\partial^2 c_i(x,t)}{\partial x^2} \tag{37}$$

In addition, the c_j ($x = 0$, $t = \infty$) term in Equation 32 has to be replaced by the c_j ($x = 0,t$) function, namely the surface activity variation of the interfering ions which can be given as follows for relatively long times $\sqrt{D't} \gg \delta$ (cf. Chapter 3, Section II and Equations 9 and 10b):

$$c_j(x = 0,t) = c_j(x,t = \infty) \times \left[1 - \frac{4}{\pi}\sum_{k=0}^{\infty}\frac{(-1)^k}{2k + 1}\exp\left(-\frac{D'(2k + 1)^2\pi^2 t}{4\delta^2}\right)\right] \quad (38)$$

if c_j $(x,t = 0) = 0$.

The primary ions adsorbed on the membrane surface are forced to desorb at a rate:

$$\frac{\partial M_i}{\partial t} = \frac{\partial M_i}{\partial c_j(x = 0,t)}\,\frac{\partial c_j(x = 0,t)}{\partial t} \quad (39)$$

parallel to the activity increase of interfering ions at the electrode surface (according to Equation 38).

The first term on the right-hand side of Equation 39 is the slope of the respective adsorption isotherm (not necessarily constant) which correlates the actual rate of desorption with the rate of change in interfering ion activity in the boundary layer. In the simplest case, the adsorption isotherm is approximately linear[261] in a wide concentration range before saturation, i.e.,

$$\frac{\partial M_i}{\partial c_j(x = 0,t)} = K \quad (40)$$

where K is the slope of the adsorption isotherm. Accordingly, by the use of Equations 36, 39, and 40, one gets:

$$-\frac{\partial M_i}{\partial t} = -K\,\frac{\partial c_j(x = 0,t)}{\partial t} = D'\,\frac{\partial c_i(x = 0,t)}{\partial x} \quad (41)$$

Thus, the more detailed description of the problem involving finite diffusion rates for both components can be described by Equations 34, 35, 37, 38, and 41, and by Equation 32 properly modified. By the use of the principle of superposition, this set of equations can be solved as follows:

$$c_i(x = 0,t) = \int_0^t c'_i(t - \tau)f(\tau)d\tau \quad (42)$$

where $c'_i(t)$ describes the concentration of primary ion generated by infinitely fast desorption and relaxation, following the stepwise increase in the interfering ion concentration, given by Equation 33 for unit amount of desorbed ions ($M_1 = 1$), while $f(\tau) \equiv \partial M_i\ (t = \tau)/\partial t$ can be given by Equation 41.

Thus, according to Equation 42, the more elaborate approach considering finite diffusion rates in both directions can be given as the mathematical convolution of the solutions of the separate diffusion problems: one of them corresponds to the extremely fast desorption of primary ions and their diffusion into the bulk of the sample solution, while the other corresponds to the diffusion of interfering ions toward the electrode surface. As only the initial period of relaxation of the desorbed elementary quantities plays a significant role in convolution, it is advantageous to use the mathematical solution for c'_i for relatively short

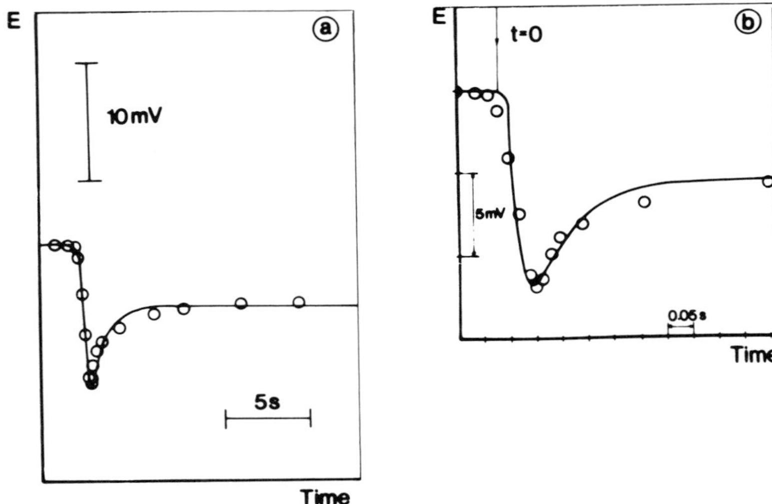

FIGURE 13. Transient signals recorded with two electrodes of different origin at 10^{-5} *M* KI activity level and constant ionic strength (10^{-1} *M* KNO_3). Solid line: theoretical curve calculated according to the more detailed model (Equation 48). Points: experimental values; experimental parameters: $c_I = 10^{-5}$ *M* KI; $c_{Br,1} = 0$. *M* KBr; $c_{Br,2} = 10^{-2}$ *M* K Br. Fitting parameters: (a) $\delta = 1.1 \times 10^{-3}$ cm, $K_{ij} = 2.05 \cdot 10^{-4}$, $K = 1.85 \cdot 10^{-5}$ cm; (b) $\delta = 8.9 \times 10^{-4}$ cm, $K_{ij} = 2.45 \cdot 10^{-4}$, $K = 3.5 \cdot 10^{-5}$ cm.

times (Equation 33). This solution converges rapidly in the time range $\sqrt{D't} \ll \delta$. In the case of $f(\tau)$, however, as the complete relaxation of the total process has to be accounted for, it is preferable to apply a solution of Equation 38, which converges rapidly at long times. Thus, Equation 33 and the derivative of Equation 38 have to be inserted into Equation 42.

Equations 42 and 38 combined with Equation 32 yield the time dependence of the electrode potential after an interfering activity step:

$$E(t) = E_i^0 - S \log \left\{ c_i^0 + \frac{K(\pi D')^{1/2} c_j(x = 0, t = \infty)}{4\delta^2 \, t^{1/2}} \int_0^t \left[1 + 2\sum_{k=1}^{\infty} (-1)^k \exp\left(- \frac{k^2\delta^2}{D'(t-\tau)} \right) \right] \left[\sum_{k=0}^{\infty} (-1)^k (2k+1) \exp\left(\frac{D'(2k+1)^2\pi^2 t}{4\delta^2} \right) \right] + K_{i,j} c_j(x = 0, t = \infty) \left[1 - \frac{4}{\pi} \sum_{k=0}^{\infty} \frac{(-1)^k}{2k+1} \exp\left(- \frac{D'(2k+1)^2\pi^2 t}{4\delta^2} \right) \right] \right\} \tag{43}$$

Equation 43 was fitted to the experimental data with the same procedure as that used in the simpler approximation described above. The result (e.g., Figure 13) is convincing proof that the earlier qualitative model (Section III.A) does not imply strongly disparate diffusion rates of the various compounds, rather the opposite is true.

The results presented in Figures 12 and 13 prove that both approaches based on the earlier model assumptions (Section III.A; the simplified solution considering extremely fast desorption, and a more comprehensive model with finite fast diffusion of both the primary and interfering ions) are suitable for the quantitative derivation of the transient signals

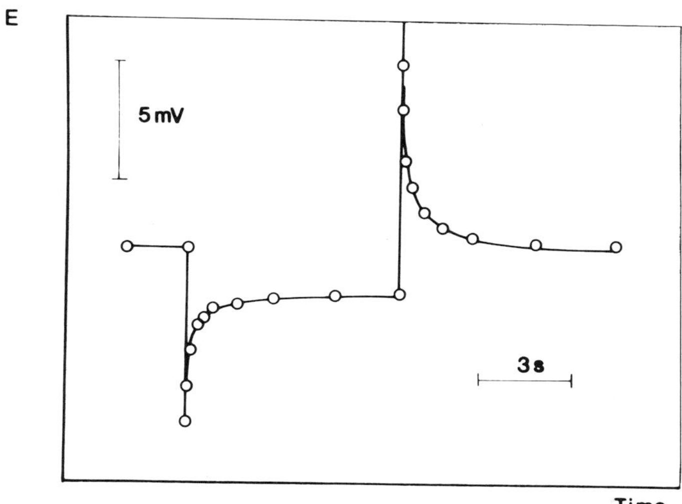

FIGURE 14. Transient signals measured (○) and calculated (solid line) with Equations 32 and 33 at 10^{-4} M KI activity level and for a bromide activity step; 0 ⇔ 10^{-1} M KBr. Fitted parameters $\delta = 8 \times 10^{-3}$ cm, $K_{i,j} = 7.9 \times 10^{-5}$, $M_i = 2.5 \times 10^{-8}$ (4.5×10^{-8}) mmol/cm^2. The fitted and $K_{i,j}$ values are identical for activity increase and decrease, but slightly different for M_i (value in parentheses).

obtained after a sudden change in the interfering ion activity. The $K_{i,j}$, M_i, and δ values determined in the course of fitting to various experimental curves are fairly adequate to the values determined by other methods.[100,261]

This treatment permits not only the description of the concentration dependence of the nonmonotonic response function (Figure 12), but also offers a simple explanation for the experimental finding that at a given primary and interfering ion activity ratio, the potential overshoot is smaller if the primary ion activity is higher (cf. Figures 12 and 14). This can be explained by the assumption that at high primary ion activities, the system is in the saturation range of the adsorption isotherm, thus the desorbed amount of primary ions cannot increase parallel to the increase of interfering ion activity.[100,261] The above model also permits the explanation of the phenomena observed when the interfering ion activity is increased in several subsequent steps (Figure 15). In such cases, the primary ions are also desorbed from the surface in several steps; consequently, several successive potential overshoots are observed in the same direction, because as long as the saturation range of the adsorption isotherm is not yet reached, the increase of interfering ion activity forces new quantities of primary ions to desorb from the electrode surface, which results in new transients.

An attempt was also made to find an experimental proof for the adsorption/desorption model assumptions. The amount of iodide ions adsorbed at the surface of AgI pellets from 10^{-5} M-labeled iodide solutions was first determined, then the pellets were placed in solutions of various bromide ion concentrations and the desorbed fraction of the previously adsorbed quantity was determined by radiochemical method (Table 2). Similar measurements were carried out with CuS-based copper (II) electrode,[102] and the amounts of copper (II) desorbed as an effect of the addition of cadmium (II) and lead (II) ions in excess were determined by atomic absorption spectrometry parallel to the recording of the nonmonotonic transient signals in the two-ion range. The experimental data obtained by both techniques supported in a quantitative way the validity of the model assumptions.

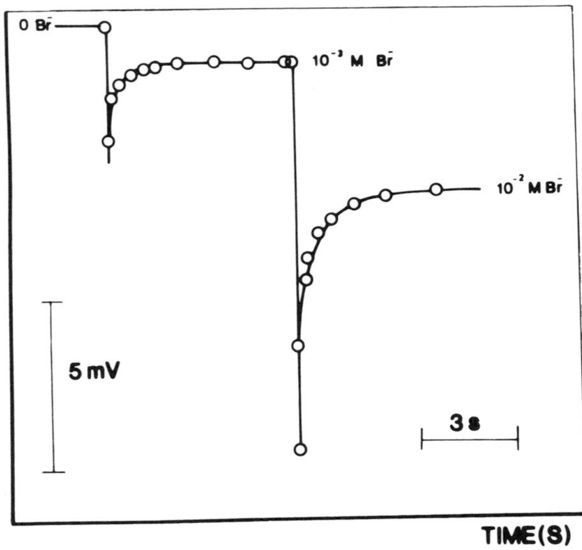

FIGURE 15. Transient signals recorded with a precipitate-based iodide electrode at the 10^{-5} *M* KI level following subsequent changes in the interfering ion activity. Solid line: theoretical curve calculated with Equations 32 and 33. Points: experimental values measured at constant ionic strength (10^{-1} *M* KNO_3). Bromide activity steps were $0 \Rightarrow 10^{-3}$ *M* KBr and 10^{-3} *M* KBr $\Rightarrow 10^{-2}$ *M* KBr. Fitted parameters were $\delta = 3.2 \times 10^{-3}$ cm, $K_{i,j} = 5.2 \times 10^{-4}$ and $M_i = 2.5 \times 10^{-9}$ mmol/cm² for $c_{Br,2} = 10^{-3}$ *M* KBr and $\delta = 3.2 \times 10^{-3}$ cm, $K_{i,j} = 2.5 \times 10^{-4}$, and $M_i = 1.8 \times 10^{-8}$ mmol/cm² for $c_{Br,2} = 10^{-2}$ *M* KBr.

IV. COMPARISON OF THE DIFFERENT MODELS

When an ISE is brought in contact with solutions containing interfering ions, the surface layer or in some cases the entire cross-section of the ion-selective membrane may reversibly or irreversibly be transformed by physical or chemical interactions.[17,49,183,191-193,209,210,252,253] Thus:

1. The concentration of the various ions in the boundary layer can differ from those prevailing in the bulk of the solution.
2. New surface layers with different properties than the bulk of the membrane can be formed. These phenomena may occur consecutively or simultaneously.

Thus, the potential of ISE is determined by both the character and kinetics of the interactions between the membrane and the interfering ions. The models (cf. Sections II and III) are aimed at the quantitative interpretation of the transient functions in the presence of interfering ions following a sudden change of the solution activity for two basically different limiting cases.

The model describing the time dependence of the apparent selectivity coefficient[183] is considered a limiting case of the segmented membrane model.[174,252] In this case, the fundamental ion-exchange process of the electrode response mechanism (Equation 14) is practically completely displaced towards the reaction products ($K_{i,j} \gg 1$). With precipitate-based electrodes, the electrode surface is transformed, while the liquid ion-exchange electrodes, the entire cross-section of the membrane is transformed during the interaction.

The transient functions observed as an effect of an interfering ion activity step are determined by the rate of these transformation reactions. The rate of the surface reactions is controlled by diffusion of the interfering ion towards the electrode surface in the case of precipitate-based electrodes (solid-state membrane approach),[183] while in the case of liquid membrane electrodes, the diffusion of the interfering ion in the membrane is the rate-determining step (liquid membrane approach).[183,251-253] The adsorption/desorption model is valid when $K_{i,j} \ll 1$; i.e., the transformation of the electrode surface is negligible even at extreme concentration ratios ($s_{eq} \approx 0$). Since the surface transformation is negligible especially at relatively short times (Figure 8: sections A and B), the potential changes are defined primarily by adsorption/desorption processes. Under such conditions, the rate of the processes is determined by the diffusion of the primary ions desorbed from the electrode surface towards the bulk of the solution, or inversely by the diffusion of the primary ions to the electrode surface from the bulk of the solution.

At a longer time after the relaxation process following the desorption and adsorption of the primary ions, the potential change due to the transformation of the electrode surface is revealed on the transient functions (Figure 8: section C; Figure 9).

The adsorption/desorption and the segmented membrane models share common features. The models are similar in many aspects. According to the segmented membrane model, the change of the apparent selectivity coefficient of the electrode in time is due to the fact that the electrode measures the activity of the primary ions liberated in a chemical reaction from the membrane, while according to the adsorption/desorption model, the primary ions exchanged and desorbed from the membrane surface are sensed by the electrode.

Both models are in formal agreement with Cammann's electrode kinetic model.[13,14] According to the latter, following the changes in interfering ion activities, the "preequilibrium current flowing through the cell (in the direction of decreasing chemical potential), necessary for the restoration of new equilibrium, can often be many times larger than that flowing under equilibrium conditions". The "preequilibrium current" can be interpreted as either the current generated in a chemical reaction of the interfering ions with the membrane material,[183,209,210] or the current produced by adsorbed or desorbed species, respectively.[180-182,184-186] The "preequilibrium current" can also be interpreted as the change of apparent selectivity coefficient in time, since the latter is defined as the contribution of each type of ions to the exchange current density according to the Cammann model.

The dynamic characteristics of ISE in the presence of interfering ions can be interpreted unequivocally by the segmented membrane model when $K_{i,j} \gg 1$,[183] and by the adsorption/desorption model when $K_{i,j} \ll 1$.[180,184] On the basis of the experimental results shown earlier (Figures 4 to 7 and 12 to 15), it can be stated that in the case of precipitate-based electrodes, if $K_{i,j} \ll 1$, the interpretation and mathematical formulation of the nonmonotonic transients by means of the time dependence of the apparent selectivity or coverage factor can be especially formal. On the other hand, the time dependence of the apparent coverage factor can be used in cases where the surface activities of primary and interfering ions are determined by membrane dissolution and new phase formation reactions. In such cases, the quantity of ions liberated or adsorbed can be neglected in comparison with that formed in the course of chemical reactions (Figure 3).

The segmented membrane model of Morf (Equations 9 to 11), on the other hand, can be used in its original form for the description of the slow or nonmonotonic signals observed in the presence of interfering ions when the ion-selective membrane consists of homogeneous layers of different properties (e.g., glass electrodes) as a result of the interaction of the membrane with the solvent or the interfering ions.

Chapter 5

DETERMINATION AND DEFINITION OF RESPONSE TIME

I. INTRODUCTION

The analytical value of terms used to assess the performance characteristic of ISE such as selectivity coefficient, lower limit of detection, response time, and life time depends on the constancy of these parameters. These parameters, however, greatly depend on the methods used for their determination.[262] Thus, these factors can be compared only if the experimental methods and conditions are carefully selected;[128,263,264] then one can expect to obtain analytically useful values only under standardized conditions.

The specification of the measuring technique and experimental conditions seems to be especially important in the case of the determination of response time, since the published values of this parameter, even for the same type of electrode, often differ by orders of magnitude.[128,134,151]

II. DYNAMIC RESPONSE OF AN ION-SELECTIVE ELECTRODE AND AN ELECTROCHEMICAL CELL

It has been mentioned in Chapter 2, Section V.B, that the transient function of a potentiometric system incorporating an ISE depends on many factors in addition to the properties of the ion-selective membrane. Thus, e.g., if the transient function of the latter is aimed to be determined in the measuring set-up shown in Chapter 2, Figure 14, then the relationship giving the time dependence of the surface ionic activity must be known in addition to the transient functions of the other elements of the system such as the electronic equipment, and other time-dependent potential sources in the measuring cell.

The transient response of the electronic unit can be determined separately by modeling the cell with a large resistance coupled to a function generator in series. Thus, the distortion of transient functions by the electronic unit can be taken into consideration. Its contribution is negligible, in general, compared with other parts of the system. Accordingly, the transient function of the measuring system can be accepted as the transient function of the electrochemical cell. Unfortunately, the response time of the indicator electrode and that of the other parts of the electrochemical cell cannot be estimated separately.[265] Accordingly, three limiting cases may be discussed:

1. The dynamic response of the electrochemical cell is primarily controlled by the measuring technique and the experimental conditions; i.e., the dynamic response of the ISE is extremely fast compared to other time-dependent processes in the cell. In most cases, the distortion of the activity step is measured as a result of the surface activity change by slow diffusion.
2. The time constants of the two parts of the electrochemical cell (Chapter 2, Figure 14) are comparable.
3. The dynamic response of the electrochemical cell is mainly determined by the properties of the indicator electrode. In this case, the activity step can be assumed to be ideal (with a rising section of infinite slope on the electrode membrane surface), and the time constant of the indicator electrode response function is much larger than that of the other parts of the electrochemical cell.

The response time of an ISE can be appropriately determined under the condition given in the third case. However, in practice, it is quite difficult to differentiate between the above limiting cases, since the types of indicator electrodes, the concentration level of primary ion, the design of the electrochemical cell, and the experimental conditions together determine the transient function of the cell. In practical applications, however, the equilibrium potential value, corresponding to the primary ion activity of the bulk, is sought, and thus only the time constant of the electrochemical cell is important.

III. DEFINITION OF RESPONSE TIME

The transient function of ISE after an activity step can be described quantitatively by a mathematical relationship incorporating the time-dependent terms or the time constant of the mathematical equation. If the latter is not available, then the rate of the potential response can conveniently be characterized by a preselected point of the transient function (response time). Thus, the transient functions can be characterized by the time necessary to attain a given percentage of the change in cell voltage due to the activity change t_α (where α indicates the percentage) or by time t* necessary to attain the new equilibrium potential with a given accuracy.

The definitions recommended by IUPAC:[263,264]

1. "The length of time (t*) which elapses between the instant at which an ISE and a reference electrode are brought into contact with a sample solution (or at which the concentration of the ion of interest in a solution on contact with an ISE and a reference electrode is changed) and the first instant at which the potential of the cell becomes equal to its steady-state value within 1 mV."[264] (Figure 1)
2. "The length of time (t_α) that elapses between the instant at which an ISE and a reference electrode are brought into contact with the sample solution (or at which the concentration of the ion of interest in a solution on contact with an ISE and a reference electrode is changed) and the first instant when the potential of the cell has reached 90% of the final value."[263] (Figure 1)

According to our view, t_α is more favorable than t*, since the latter is different for ISE measuring mono- or divalent ions even if the time constants of the respective transient functions are equal (see Figure 1; Chapter 3, Equation 22).

Uemasu and Umezawa[163] pointed out the paradox involved in these definitions, namely, that one cannot determine t_α and t* without the knowledge of the equilibrium potential (E_2) corresponding to the stepped activity value: i.e., the above definitions are not of much practical use.

Lindner et al.[6,128,165,222] showed that the transient functions of different types of ISE could be compared easily by the slope values (m) evaluated at a given point (e.g., starting or inflexion point) of the transient response curve. The normalized slopes (m_{eff}) can be used for the comparison of transient functions of mono- or divalent ISE or of electrodes exhibiting nonideal (non-Nernstian) potential response. It is also helpful for comparing transient functions recorded at different activity ratios (see Chapter 2, Figure 20; Chapter 3, Figure 15).

The technique of the evaluation of m data is shown in Figure 2, and the normalized slope m_{eff} (mV/msec) is given by Equation 1:

$$m_{eff} = m(t) \cdot \frac{\Delta E^x}{\Delta E} \qquad (1)$$

where m(t) is the experimentally determined slope at a given time instant (mV/msec) (Figure 2b); ΔE is the measured change in cell voltage corresponding to the activity change of the sample solution ($\Delta E = E_2 - E_1$) (mV); and ΔE^x is the theoretical potential change corresponding to the one-decade activity change (5 g 16 mV at 25°C).[59]

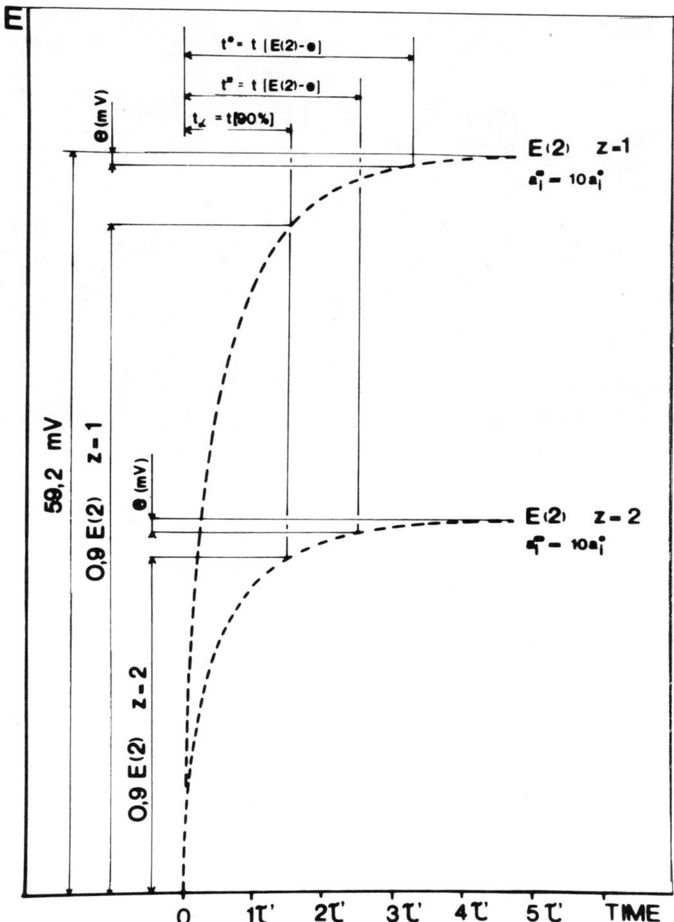

FIGURE 1. Theoretical response time curves of uni- and divalent ISE calculated with Equation 19 in Chapter 3. t_α ($t_{90\%}$) is the time required to attain 90% of the final (steady-state) potential E_2; t* ($t_{E_2 - e}$) is the time required to attain the steady-state potential value (E_2) within ± e(mV).

It was shown experimentally that at relatively high sample flow rates, the initial slopes (or the normalized ones) of the transient functions were not exclusively determined by the diffusion process across the adhering solution film; i.e., they may provide some kinetic information about processes following the ion transport in the solution (Chapter 3, Table 2).[152] Since the rising section of the transient functions often have a zero initial slope (Chapter 3, Figure 21) and the transient functions generally consist of several sections (e.g., Chapter 3, Figure 19), the determination of the slope value at the inflection point is suggested (Figure 2b).

On the basis of favorable experiences gained in comparing transient functions of ISE with m_{eff}[6,128,165,222] Uemasu and Umezawa[163] defined a value called differential quotient, t(ΔE/Δt), as a measure of practical response time. According to the authors' definition, the transient functions can be characterized with the time elapsed between the instant at which an ISE and a reference electrode (electrochemical cell) are brought into contact with a sample solution and the first instant at which the transient function reaches a predetermined slope (differential quotient, ΔE/Δt) (Figure 3).

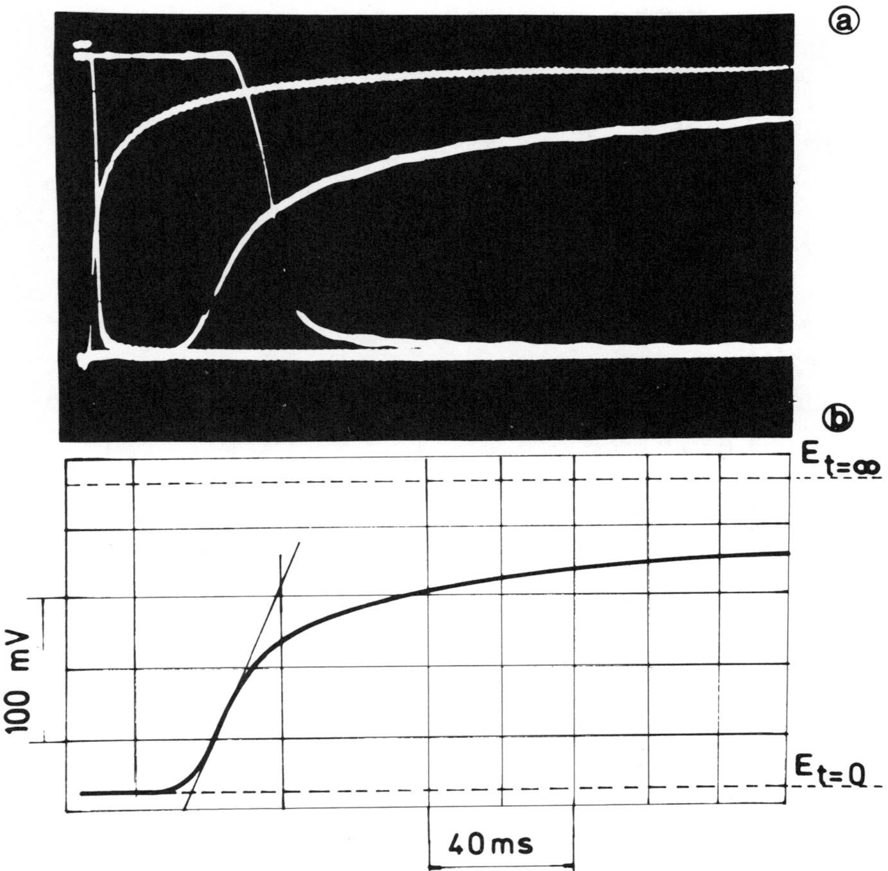

FIGURE 2. (a) Experimental response time curves recorded at different resolutions with respect to time: $a_{i,1}$ = 10^{-1} *M* KI + 10^{-2} *M* KNO_3; $a_{i,2}$ = 10^{-5} *M* KI + 10^{-1} *M* KNO_3; or vice versa. y = 50 mV/division; x = 20 or 200 msec/division. (b) The evaluation of "initial" slopes using as an example the curve shown in (a): $a_{i,1}$ = 10^{-5} *M* KI + 10^{-1} *M* KNO_3; $a_{i,2}$ = 10^{-1} *M* KI + 10^{-2} *M* KNO_3. y = 50 mV/division; x = 20 msec/division.

Several problems related to the accepted definitions of response time (t_α or t*) can be overcome by the introduction of the above new definition which has the following advantages:

1. It can be determined easily.
2. It can be determined without knowing the E_2 value.
3. It has the same significance as any other response time definition.
4. It can also be defined in the case of response time curves consisting of sections corresponding to different rate-determining processes.
5. It is of practical importance since it helps the analyst to define the time of the potential reading during potentiometric analysis.[266]

For practical analytical purposes, such slope (differential quotient) values should be selected which fall on the asymptotic part of the transient functions.[163,228] Since response time data dramatically increase with the decrease of the preselected differential quotient, the practical differential quotient values must be selected according to the required analytical accuracy.

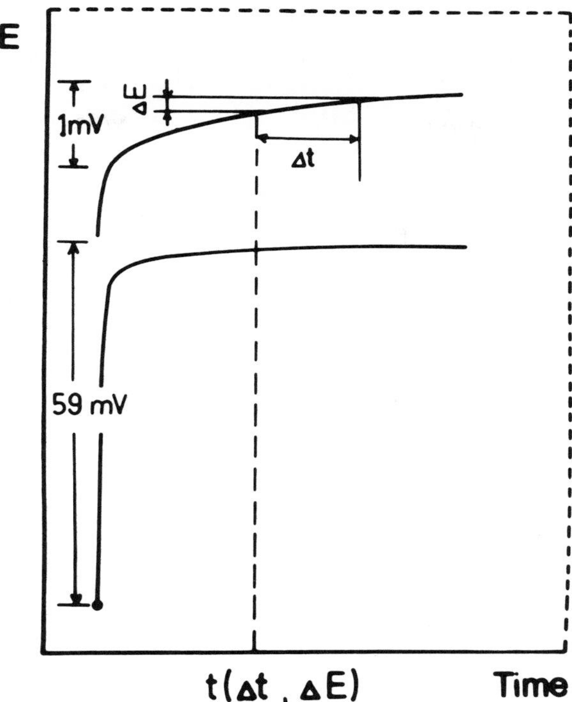

FIGURE 3. Definition and determination of t(ΔE/Δt).

In addition to the above-mentioned advantages, the new slope-based definition also has drawbacks.[128,228] If the potential data corresponding to the preselected differential quotient value are used instead of the equilibrium potential (E_2), it is rather difficult to predict the error of the analytical determination. The error depends on the magnitude of the activity step ($a_{i,1}/a_{i,2}$); on the slope of the electrode response function (E vs. log a_i); as well as on the conditions of the electrode membrane. However, these drawbacks also apply to earlier definitions of response time.[263,264] In order to eliminate the errors in readings, a so-called slope controller, calculating the slope of response time curves at given time instants, is built in some pH meters manufactured by Radelkis.[266] Thus, the potential value corresponding to a critical slope ($\Delta E/\Delta t$) is considered as an equilibrium value. However, the advantages of this slope controller can be questioned as far as the error of the analytical determination is concerned, but it certainly has an advantage in cases of a known constant drift in the cell potential.

A comparison of response time data determined according to different mathematical models and response time definitions is in Tables 1 and 2.

The t_α and t^* values were calculated with the following formulas derived on the basis of the diffusion of the primary ion in either the stagnant solution layer (cf. Chapter 3, Section II and Equation 22) or in the ion-selective membrane (cf. Equation 2; Chapter 3, Section IV), respectively:

$$t(\epsilon) = \tau' \ln \frac{1 - (a_{i,1}/a_{i,2})}{1 - (a_{i,1}/a_{i,2})^{\epsilon}} \tag{2}$$

$$t(\epsilon) = \tau \left[\frac{(a_{i,1}/a_{i,2})^{\epsilon} - (a_{i,1}/a_{i,2})}{1 - (a_{i,1}/a_{i,2})^{\epsilon}2} \right] \tag{3}$$

Table 1
THEORETICAL RESPONSE TIME VALUES CALCULATED WITH EQUATIONS 19 AND 22 IN CHAPTER 3 (VALUES IN SECONDS, CALCULATED WITH $\tau' = 1$ sec)[a]

	t_α [b]			t^* [b]			$t(\Delta E/\Delta t)$		
$a_{i,2}/a_{i,1}$	50%	90%	99.5%	1 mV	0.5 mV	0.1 mV	1 mV/min	0.5 mV/min	0.1 mV/min
Activity increase									
10	0.27	1.47	4.36	3.16	3.84	5.44	8.07	8.76	10.37
100	0.09	1.47	3.77	3.25	3.93	5.54	8.16	8.88	10.47
Activity decrease									
0.1	1.43	3.41	6.66	5.42	6.12	7.74	10.37	11.06	12.67
0.01	2.39	5.13	8.36	7.82	8.52	10.14	12.77	13.46	15.07

[a] Time constant value used for practical analytical conditions.[7,142]
[b] In the case of t_α and t^*, the response time data increase proportional with τ'.

Table 2
THEORETICAL RESPONSE TIME VALUES CALCULATED WITH EQUATION 2 (AND EQUATION 88 (CHAPTER 3) (VALUES IN SECONDS, CALCULATED WITH $\tau = 10$ msec[a])

	t_α [b]			t^* [b]			$t(\Delta E/\Delta t)$		
$a_{i,2}/a_{i,1}$	50%	90%	99.5%	1 mV	0.5 mV	0.1 mV	1 mV/min	0.5 mV/min	0.1 mV/min
Activity increase									
10	10^{-3}	0.11	60.3	5.1	20.9	532	29.1	46.2	136
100	10^{-4}	0.03	18.1	6.2	25.4	644	31.0	49.3	145
Activity decrease									
0.1	0.1	11.4	6025	509	$2.1 \cdot 10^3$	$5.3 \cdot 10^4$	128	206	617
0.01	1.0	283	$1.8 \cdot 10^5$	6.10^4	$2.5 \cdot 10^5$	$6.4 \cdot 10^6$	530	886	2793

[a] Time constants measured with neutral carrier electrodes.[7,222]
[b] In the case of t_α and t^*, the response time data are proportional with τ.

where $t(\epsilon)$ is the response time, while (Chapter 3, Equation 23):

$$\epsilon = \frac{E(t) - E_2}{E_1 - E_2} \tag{4}$$

These equations permit the calculation of both t_α and t^*. The derivatives of Equations 19 and 88 in Chapter 3 were used for the calculation of $t(\Delta E/\Delta t)$ values.

It is apparent from the foregoing discussion that a generally applicable procedure for the determination of the response time is conspicuous by its absence. However, the use of the "differential quotient"-based definition is recommended from a practical point of view as the most satisfactory approach for the time being.[128]

IV. ESTIMATION OF THE EQUILIBRIUM POTENTIAL

It is clear from the preceding review of the attempts to define response time that a meaningful and universally acceptable definition would require a complete mathematical description incorporating all experimental parameters affecting the transient function of potentiometric cells. Although such a general mathematical equation does not exist, the mathematical models derived so far for the description of the potential time curves permit the calculation of steady-state or equilibrium potential values only in some well-defined limiting cases:

1. The dynamic response of the potentiometric system (Chapter 2, Figure 14) is mainly controlled by diffusion through the adhering solution film. In general, this is the case in the absence of interfering ions when glass-, precipitate-, or liquid ion-exchanger-based electrodes are used in primary ion concentrations higher than 10^{-4} *M* under practical analytical conditions: e.g., in cells without mixing, in slightly stirred solutions, or in flow-through cells at low flow rates (slow measuring setup). Under these conditions, the diffusion through the adhering layer is the rate-controlling process, and the transient functions can be described by Chapter 3, Equation 19 (see also Chapter 3, Figure 5).
2. The dynamic response of the system is controlled by the transient function of the measuring electronics (Chapter 2, Figure 14). This can occur when electrodes having extremely high resistance are used (e.g., microelectrodes). In such cases, Equation 8 in Chapter 3 is suggested for expressing the transient function of the system.
3. The dynamic response of the electrochemical cell is mainly controlled by the properties of the indicator electrode (electrodes with slow response time, specially designed cell, and fast measuring electronics): (1) transient functions of cells with precipitate-based electrodes in the range of the detection limit. Under these conditions, the experimentally recorded transient function can be described by a hyperbolic equation (Chapter 3, Equations 70 to 74 and Figure 24); (2) transient functions of cells containing neutral carrier-based ISE. The transient functions can be described by assuming the diffusion within the ion-selective membrane as the rate-controlling process (Chapter 3, Equation 88).

In practice, in the above limiting cases, the respective mathematical equation can be fitted to the experimental data with the required accuracy (Figure 4; Chapter 3, Figures 5, 23, and 24) and it is immaterial which of the current definitions is used for the determinations of response time (Figures 1 and 3).

The determination of the equilibrium electrode potential by extrapolation is necessary mainly when the transient function of the electrode approaches the steady-state value slowly. Such behavior is typical with ionophore-based electrodes. The transient functions of this type of electrodes depend on the square root of time (Figure 3; Chapter 3, Equation 88 and Figure 27); thus, $t_{99.9} \approx 100\ t_{99}$.

On the contrary, with ISE with constant membrane composition (Chapter 3, Equation 19) $t_{99.9} \approx 3t_{90}$, there is no particular need to use the extrapolation technique for the estimation of the steady-state cell voltage.

With the use of special analytical methods (e.g., flow-through techniques which employ ISE-based potentiometric detection), it would be advantageous if the equilibrium electrode potential (E_2) were determined with high accuracy from a given ($E(t_1) - E(t_2)$) section of the transient function. Simon and co-workers[267] obtained remarkably good results with microprocessor-controlled equipment developed for the measurement of the ionic constituents of blood serum using nonlinear regression and Equation 88 in Chapter 3 in the calculation of the equilibrium electrode potential (Figure 4).

The dynamic properties of ISE (or of the potentiometric system) among others can also be influenced by the nature of the sample. Under these conditions, accurate analytical results can be expected only if the evaluation is based on equilibrium potential (E_2) data. If the

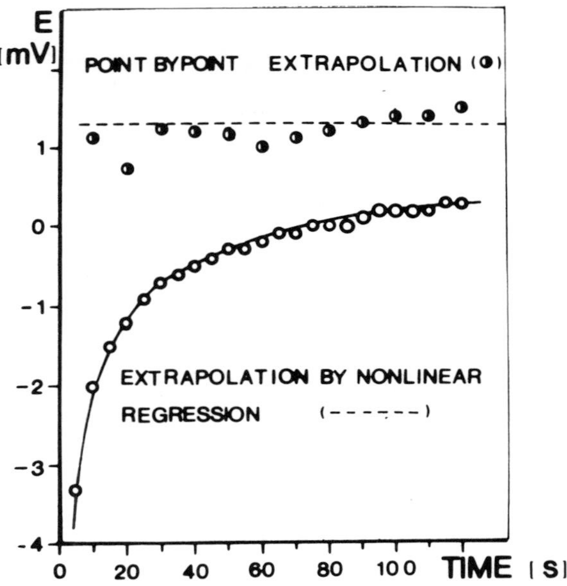

FIGURE 4. Determination of the steady-state EMF of a valinomycin-based membrane electrode (slowly stirred sample) using different extrapolation procedures. Nonlinear regression curve fit by Equation 88 in Chapter 3. Point-by-point extrapolation: early stage estimation inserting $E_2 = 3.414\ E(t) - 2.414\ E(t/2)$. (From Morf, W. E., *The Principles of Ion-Selective Electrodes and of Membrane Transport*. Akadémiai Kiadó, Budapest, 1981, With permission.)

evaluation of analytical results, however, is based on a transient potential signal [E(t)] as in flow injection techniques,[268-270] then a decrease in precision can hardly be excluded.

The equilibrium electrode potential can be calculated by curve fitting if the mathematical relations (or their linearized form) are known from a given initial section of the transient function (Figure 4; Chapter 3, Figures 23 and 24).

V. SELECTED EXAMPLES

It was originally one of the aims of this book to give a survey of the response time data published in the literature for different types of ISE. However, as it follows from the foregoing chapters, the simple tabulation of the published response time data can be rather contradictory since the data differ by orders of magnitudes even for the same type of electrode. Light and Schwartz[133] as well as Denks and Neeb[146,160,161] measured the response time of Ag_2S-based silver ISE and reported values in the millisecond range, while Hseu and Rechnitz[127] and later Thomas[271] published values of some seconds, and finally Shatkay[134] published response time data of several hours. In the light of the considerations discussed in the foregoing chapters the contradictions are often apparent, which is intended to be shown on some selected examples taken from the literature.

Response times ranging from a few milliseconds to some minutes were published for silver halide-based electrodes of the second kind,[171] for liquid ion-exchanger-based nitrate[170] and calcium[139,142] ISE, for cation-selective glass electrodes[144] and pH-sensitive glass electrodes,[115,142] as well as for various types of precipitate-based electrodes.[4-8,127,133,141,145,146,148,154,160,161,169]

Table 3
COMPARISON OF $t_{50\%}$, $t_{90\%}$, and $t_{95\%}$ DATA CALCULATED ACCORDING TO THE DIFFERENT MATHEMATICAL MODELS IN CHAPTER 3

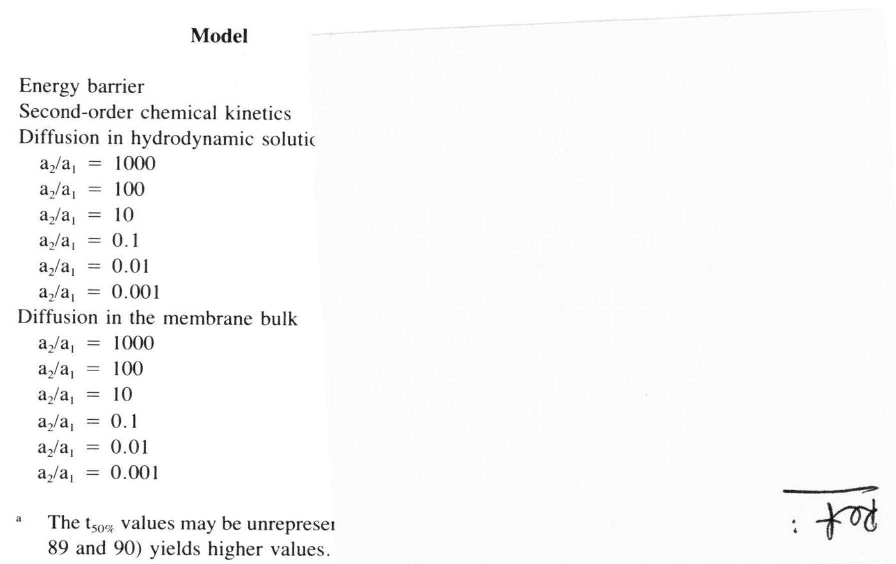

Model
Energy barrier
Second-order chemical kinetics
Diffusion in hydrodynamic solutic
a_2/a_1 = 1000
a_2/a_1 = 100
a_2/a_1 = 10
a_2/a_1 = 0.1
a_2/a_1 = 0.01
a_2/a_1 = 0.001
Diffusion in the membrane bulk
a_2/a_1 = 1000
a_2/a_1 = 100
a_2/a_1 = 10
a_2/a_1 = 0.1
a_2/a_1 = 0.01
a_2/a_1 = 0.001

[a] The $t_{50\%}$ values may be unrepreseı
89 and 90) yields higher values.

There are reports of response time data of some minutes, e.g., ionophore-based sodium[167] and potassium[7] ISE, pH[95] and sodium ion[167]-selective glass electrodes, and various types of precipitate-based electrodes.[24,25,125,126,134,158,159-166] Moreover, response times of several hours can also be found (e.g., for cation-selective glass electrodes[123,124] and various precipitate-based electrodes.[134,136,156,157]

These very different data and time constants can be explained partly by the discrepancies in the definition of the response time. This is demonstrated by data in Tables 1 and 2, calculated according to two different theoretical models (cf. Chapter 3, Sections II and IV).

On the other hand, the measured response time data corresponding to different definitions can be interrelated if the respective mathematical models are known. The calculations of such response time values are easier if the ratios of the most frequently used $t_{50\%}$, $t_{90\%}$, and $t_{95\%}$ data are known for the different model assumptions (Table 3). Moreover, if the complete transient function is known and the ratio of $t_{50\%}$:$t_{90\%}$:$t_{99\%}$ is calculated, then the data in Table 3 may permit the selection of the most suitable model for the description of the transient function.

Before demonstrating the applicability of Table 3 on some selected examples, the following general conclusions are drawn:

1. Response times in the millisecond range were almost exclusively obtained with setups specially designed for response time measurements allowing high flow or high stirring velocities.[6,8,141,144,146] Thus, the change in the activity of the sample solution on the electrode surface can be approximated by an ideal step function. Under these conditions, the distorting effect of ion diffusion through the hydrodynamic boundary layer, in general, can be neglected; i.e., the transient function of the electrochemical cell is mainly determined by the properties of the indicator electrode.
2. Response times in the range of some minutes or hours were exclusively observed in either very dilute solutions or with electrodes (lanthanum fluoride-based fluoride-sensitive electrode; glass electrode) of high resistance surface layer.[24,25,123,138,152,156]

Table 4
COMPARISON OF THE DYNAMIC RESPONSE OF A FLUORIDE-SELECTIVE ELECTRODE (ELECTRODE B[a]) BASED ON $t_{90\%}$ AND $t_{50\%}$ DATA DETERMINED IN TWO DIFFERENT SETUPS BY MERTENS ET AL.[141]

	$t_{90\%}/t_{50\%}$ determined in	
F^- concentration setp (M)	Perspex conical vacuum chamber (see Chapter 2, Figure 17): flow rate = 7.5 mℓ/sec	Automatic potentiometric system (see Reference 141, Figure 3): flow rate = 1 mℓ/min
$10^{-5} \rightarrow 10^{-4}$	9.0	3.9
$10^{-4} \rightarrow 10^{-3}$	9.28	4.54
$10^{-3} \rightarrow 10^{-2}$	6.66	3.4
$10^{-4} \rightarrow 10^{-5}$	10.0	3.03
$10^{-3} \rightarrow 10^{-4}$	10.9	3.56
$10^{-2} \rightarrow 10^{-3}$	9.5	3.67
Mean	9.22	3.68
SD	1.42	0.51

[a] Orion 94-09A fluoride-selective electrode (2.5 years old).

From Mertens, J. and Van den Wirkel, P., *Anal. Chem.*, 48, 272, 1976. With permission.

3. The response time can increase considerably in the presence of interfering ions. Thus, the transient response function can be nonmonotonic if the interfering ion activity is changed stepwise during the measurement. The response time definitions derived for monotonous transient functions are obviously inapplicable for nonmonotonic transients.

A. Interpretation of the Data of Mertens Et Al.

Mertens et al.[141] investigated the dynamic response of the fluoride ion-sensitive electrode by a fast flow technique in a specially designed apparatus (shown in Chapter 2, Figure 17) and in a flow-through automatic system. The authors obviously found a discrepancy between the dynamic response measured in the two different equipments. A hyperbolic function was found to be appropriate to describe the transient functions measured in the "vacuum conical chamber" (Chapter 2, Figure 17; Chapter 3, Equation 75 and Figure 23). On the contrary, the film diffusion across the electrode contacting solution layer is assumed to be the rate-controlling process on the basis of the flow rate dependence of the transient signals recorded in the autoanalyzer-type flow system. Both statements can be confirmed simply by comparing the data shown in Tables 3 and 4.

These results prove that a drastic change in cell geometry and experimental conditions (the flow rate was 7.5 mℓ/sec in the first case, while only 1 mℓ/min in the second) affects the actual value of the response time, but, moreover, may alter the rate-controlling step in the overall electrode reaction, i.e., the mathematical function used to describe the transient function.

B. Interpretation of the Data of Denks and Neeb

Denks and Neeb[146,160,161] compared the dynamic properties of metallic silver and Ag_2S/AgI-based silver and iodide ISE in a favorably designed experimental cell by the activity step method (Chapter 2, Figure 18). The transient functions observed were characterized by $t_{50\%}$, $t_{90\%}$, and $t_{95\%}$ (Tables 5 to 7).

The shortest response times were observed for silver metal electrodes,[146] in accordance with earlier findings by Lindner et al.[6] The latter found that the response times of Ag_2S/AgX-based electrodes were shorter for silver ion than for the corresponding halide ions.

Table 5
RESPONSE TIME DATA OF A SILVER METAL ELECTRODE IN SOLUTIONS CONTAINING SILVER IONS (I = 0.1; $NaNO_3$)

No.	Range	$t_{50\%}$ (msec)	$t_{90\%}$ (msec)	$t_{95\%}$	$t_{50\%}:t_{90\%}:t_{95\%}$	$t_{95\%}:t_{90\%}$
1	$10^{-5}/10^{-4}$	2.6 (8.0)	10 (40)	13 (90)	1:3.85 (5):5 (11.3)	1.3 (2.3)
2	$10^{-5}/10^{-3}$	1.8 (8.0)	6 (60)	8 (140)	1:3.3 (7.5):4.44 (17.5)	1.3 (2.3)
3	$10^{-5}/10^{-2}$	1.4 (9.0)	5 (110)	7 (250)	1:3.57 (12.2):5 (27.8)	1.4 (2.3)
4	$10^{-4}/10^{-3}$	1.8 (3.3)	6 (12)	8 (18)	1:3.3 (3.6):4.44 (5.45)	1.3 (1.5)
5	$10^{-4}/10^{-2}$	1.4 (5.4)	5 (26)	7 (55)	1:3.57 (4.8):5 (10.2)	1.4 (2.11)
6	$10^{-4}/10^{-1}$	1.4 (9.5)	5 (90)	7 (280)	1:3.57 (1.5):5 (29.5)	1.4 (3.11)
7	$10^{-3}/10^{-2}$	1.8 (2.8)	6 (9)	8 (14)	1:3.3 (3.2):4.44 (5)	1.3 (1.55)
8	$10^{-3}/10^{-1}$	1.5 (6.0)	5 (28)	7 (65)	1:3.3 (4.6):4.67 (10.8)	1.4 (2.3)
9	$10^{-2}/10^{-1}$	1.8 (2.8)	6 (9)	8 (13)	1:3.3 (3.2):4.44 (4.6)	1.3 (1.4)
				Mean	1:3.54 (5.95):4.71 (13.57)	1.3 (2.09)

Note: Data in parentheses are for activity decrease.

From Denks, A., Neeb, R., and Fresenius, Z., *Anal. Chem.*, 285, 233, 1977. With permission.

Table 6
RESPONSE TIME DATA OF AN Ag_2S/AgI ELECTRODE IN IODIDE SOLUTIONS (I = 0.1; $NaNO_3$)

No.	Range	$t_{50\%}$ (msec)	$t_{90\%}$ (msec)	$t_{95\%}$ (msec)	$t_{50\%}:t_{90\%}:t_{95\%}$	$t_{95\%}:t_{90\%}$
1	$10^{-5}/10^{-4}$	7.2 (19)	45 (200)	85 (480)	1:6.25 (10.52):11.8 (25.26)	1.9 (2.4)
2	$10^{-5}/10^{-3}$	3.2 (14)	20 (270)	42 (740)	1:6.25 (19.3):13.12 (52.85)	2.1 (2.7)
3	$10^{-5}/10^{-2}$	2.0 (18)	17 (550)	35 (1500)	1:8.5 (30.5):17.5 (83.33)	2.1 (2.7)
4	$10^{-4}/10^{-3}$	3.5 (7.6)	24 (70)	47 (250)	1:6.85 (9.21):13.42 (32.89)	2.0 (3.6)
5	$10^{-4}/10^{-2}$	2.3 (9.7)	23 (160)	54 (510)	1:10 (16.49):23.47 (52.57)	2.4 (3.2)
6	$10^{-4}/10^{-1}$	2.4 (19)	18 (730)	37 (2800)	1:7.5 (38.42):15.41 (147.36)	2.1 (3.8)
7	$10^{-3}/10^{-2}$	3.1 (7.0)	40 (185)	130 (630)	1:12.9 (26.42):41.93 (90)	3.3 (3.4)
8	$10^{-3}/10^{-1}$	2.7 (11)	25 (300)	50 (1300)	1:9.25 (27.27):18.5 (118.18)	2.0 (4.3)
9	$10^{-2}/10^{-1}$	3.3 (7.5)	45 (250)	130 (1500)	1:13.63 (33.33):39.4 (200)	2.9 (6.0)
				Mean	1:9.01:21.61	2.31

Note: Data in parentheses are for activity increase.

From Denks, A., Neeb, R., and Fresenius, Z., *Anal. Chem.*, 297, 121, 1979. With permission.

It is apparent from a comparison of Tables 3 and 5 that the transient function of a metallic silver electrode can closely be approximated by an exponential-type equation such as Equation 19 or 54 in Chapter 3, or a linearized form of the former (Equation 21) which is analogous with Equation 54. On the contrary, the transient functions of a Ag_2S/AgI-based iodide electrode can be described with sufficient accuracy by a hyperbolic relationship (cf. Tables 3 and 6) such as, e.g., Equations 70 and 72 to 74 in Chapter 3. However, the transient functions of a Ag_2S/AgI-based electrode in silver ion-containing solution fall between the two limiting cases.

The above experimental results can be interpreted by supposing that diffusion through the hydrodynamic boundary layer is the rate-determining step in the case of the silver metal electrode having the shortest response time (Chapter 3, Equations 19 and 21). Thus, the "clear" exponential form of the transient functions recorded after an activity increase is

Table 7
RESPONSE TIME DATA OF AN Ag_2S/AgI ELECTRODE IN SOLUTIONS CONTAINING SILVER IONS (I = 0.1; $NaNO_3$)

No.	Range	$t_{50\%}$ (msec)	$t_{90\%}$ (msec)	$t_{95\%}$ (msec)	$t_{50\%}:t_{90\%}:t_{95\%}$	$t_{95\%}:t_{90\%}$
1	$10^{-5}/10^{-4}$	2.2 (6.8)	9 (46)	13 (80)	1:4.09 (6.76):5.9 (11.76)	1.44 (1.73)
2	$10^{-5}/10^{-3}$	2.1 (10)	9 (78)	20 (160)	1:4.28 (7.8):9.52 (16)	2.22 (2.05)
3	$10^{-5}/10^{-2}$	1.5 (12)	6 (190)	9 (390)	1:4.0 (15.83):6 (32.5)	1.5 (2.05)
4	$10^{-4}/10^{-3}$	1.9 (4.5)	7 (24)	12 (50)	1:3.68 (5.33):6.31 (11.11)	1.71 (2.08)
5	$10^{-4}/10^{-2}$	1.8 (7.0)	7 (47)	13 (100)	1:3.88 (6.71):7.22 (14.28)	1.85 (2.12)
6	$10^{-4}/10^{-1}$	1.6 (10)	7 (140)	14 (360)	1:4.37 (14):8.75 (36)	2.0 (2.57)
7	$10^{-3}/10^{-2}$	2.2 (4.8)	15 (25)	22 (45)	1:6.81 (5.21):10 (9.375)	1.46 (1.8)
8	$10^{-3}/10^{-1}$	2.8 (7.5)	13 (48)	22 (100)	1:4.64 (6.4):7.85 (13.33)	1.69 (2.08)
9	$10^{-2}/10^{-1}$	2.6 (4.9)	16 (27)	31 (50)	1:6.15 (5.5):11.92(10.20)	1.93 (1.85)
				Mean	1:4.65:8.16	1.75

Note: Data in parentheses are for activity increase.

From Denks, A., Neeb, R., and Fresenius, Z., *Anal. Chem.*, 285, 233, 1977. With permission.

readily understood (cf. Tables 3 and 5). Deviations from the diffusion model were encountered parallel to an increase in response time in relatively dilute solutions and when activity steps larger than one order of magnitude were introduced (cf. Figures 14 and 15 in Chapter 3). This phenomenon was discussed in detail in Chapter 3, Section II.A.

In accordance with the above observations, the ratios of the $t_{50\%}$, $t_{90\%}$, and $t_{95\%}$ data found experimentally at an activity decrease may differ considerably from the theoretical values corresponding to exponential functions (Table 3). In the case of activity decrease, only the measurements performed at one-decade activity change and at relatively concentrated solutions (Table 5: nos. 4, 7, and 9; i.e., if the $t_{95\%}$ was found less than 20 msec) correspond to the models described with exponential functions (Chapter 3, Equations 19, 21, and 54).

It is apparent from the comparison of the data (especially that of $t_{95\%}$ values) of Tables 5 and 6 that the response time of Ag_2S/AgI-based iodide electrode is 5 to 15 times longer than that of the silver metal electrode. Thus, the transient function of the electrochemical cell in the given experimental setup is determined by the properties of the precipitate-based iodide electrode. The rate of diffusion in the hydrodynamic boundary layer was found to be negligible compared to other partial reactions of the electrode process. The transient function was found to have a hyperbolic form in accordance with the literature data.[24,25,141] The ratios of times $t_{50\%}$, $t_{90\%}$, and $t_{95\%}$ are equal to those given in Table 3 for hyperbolic functions. However, the response time of Ag_2S/AgI-based ISE at silver ion response was found to be between that of a silver metal and a Ag_2S/AgI-based iodide electrode (cf. Tables 5 to 7). This fact suggests that the diffusion through the phase solution boundary layer and the other steps of the electrode processes has commensurable rates. The shape of the transient functions is determined by both processes depending on the relative values of the rates of the partial processes. Thus, the respective ratios of $t_{50\%}$, $t_{90\%}$, and $t_{95\%}$ are between the values characteristic for exponential and hyperbolic functions, respectively.

Denks and Neeb[146,160,161] rejected the model based on diffusion control through the hydrodynamic boundary layer even in the case of silver metal electrode under the special experimental conditions prevailing in the specially designed measurements (large linear flow velocity; minimum thickness of hydrodynamic boundary layer). Accordingly, the exponential character of the transient functions of silver metal electrode can be explained on the basis of the energy barrier model (Chapter 3, Section III.A and Equations 54 and 55) and the assumption of first-order kinetics (Chapter 3, Section III.B and Equation 66).

Under such experimental conditions, the rate of diffusion through the hydrodynamic boundary layer is negligible compared to the other electrode processes even in the case of silver metal electrodes. This statement is also valid for Ag_2S/AgI-based iodide electrodes which exhibit much longer response times than metallic silver. Thus, the transient functions of the iodide electrode can be described with hyperbolic expressions.[24,25,141] In such conditions, the transient functions of Ag_2S/AgI-based electrodes in solutions containing silver ions are also interpreted with the assumption of mixed kinetics (e.g., Chapter 3, Section III.B), except that diffusion through the boundary layer does not affect the rate of the overall electrode reaction.

Chapter 6

THE IMPORTANCE OF DYNAMIC PROPERTIES OF ION-SELECTIVE ELECTRODES IN PRACTICAL ANALYSIS

The potentiometric sensors are frequently employed as sensors in different analytical methods including flow techniques.[244-248,270] Ion-selective electrodes (ISE) are, e.g., reliable tools for the high precision monitoring of ion activities in biological samples in continuous flow, or *in situ* applications.

For the optimalization of conditions during the development of an analytical method based on potentiometric detection, the most important characteristics of the potentiometric sensors must be considered, among which the dynamic properties of the sensors are of utmost importance. The latter, i.e., the dynamic response characteristics of the sensors or cells, can drastically affect not only the attainable rate of analysis, but also the magnitude of the signal measured and thus the sensitivity of the measurements in batch, continuous flow, and flow injection analysis (Figure 1).

With the aim of increasing the rate and precision of measurements in continuous flow systems based on neutral carrier ISE, Morf and Simon[17,132] and Deutsch et al.[267] suggested a method for determining the steady-state potential from the rising part of a dynamic response curve by the use of an appropriate mathematical equation and extrapolation by nonlinear regression analysis. This method is of utmost importance if the final value of the steady-state electrode potential is approached slowly.

In flow injection systems based on sensors with a finite response time, the sensitivity of the measurements can be increased theoretically by increasing the residence time of the sample in the detector cell (Figure 2). This can be achieved by decreasing the flow rate or increasing the sample volume. By decreasing the flow rate, the sampling frequency will decrease, but at the same time, the response time is expected to be longer and, accordingly, the increase in sensitivity cannot be attained in general. Residence time can also be increased, but at the cost of sample dilution by a mixing chamber of appropriate volume or a larger reactor volume.

The sensitivity of the measurements is expected to be increased by the increase of sample volume injected (Figure 1). The amount of material injected, however, can be kept constant if the injected sample is proportionally diluted before injections, or it can be increased if the sample concentration is unchanged. In the former case, the decrease of the concentration of the sample injected may result in an increase of the response time, especially if the diluted sample concentration will fall in the range of the detection limit of the sensor.[152] Thus, as a consequence, the sensitivity and the rate of analysis may become less favorable. On the other hand, applying an increased sample volume at constant concentration, a loss in sensitivity is attained if the sensitivity is defined by the minimum amount of the component to be determined. However, if the sensitivity is defined by the minimum concentration that can be determined, the increase of sample volume can be regarded as a gain in sensitivity.

In Figure 3, the transient response of the potentiometric cell in flow injection analysis is shown. In case A, the concentration of the streaming background solution is smaller than that of the injected sample. In case B, the opposite is true.

As it was shown earlier (see Chapter 2, Section V.C.1 and C.3), the response time is shorter if the activity of the sample solution is increased. Accordingly, by injecting very small sample volumes, higher sensitivity is attained in case A, but the time needed for one analysis increases compared to case B. On the other hand, it is apparent in Figure 3 that the sampling frequency can be higher in case B than in case A, because the decay of the signal to the base line is expected to be faster. The application of a streaming background

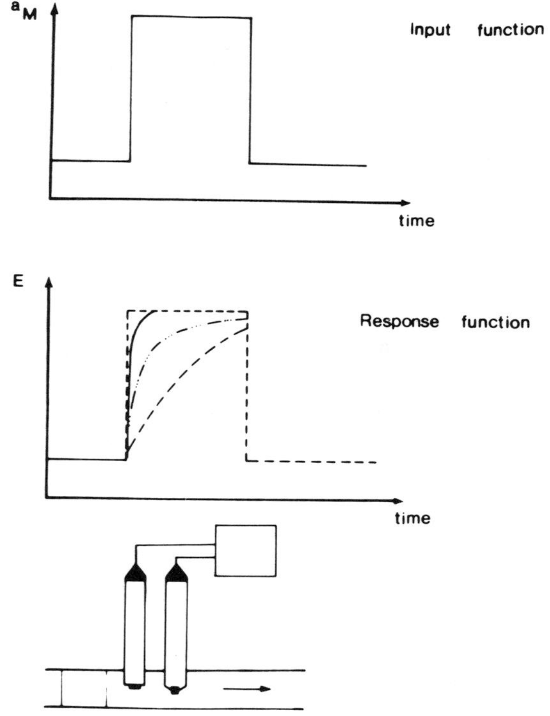

FIGURE 1. Schematic illustration of the effect of the dynamic characteristics of the potentiometric sensor on the potential recorded in flow injection analysis.

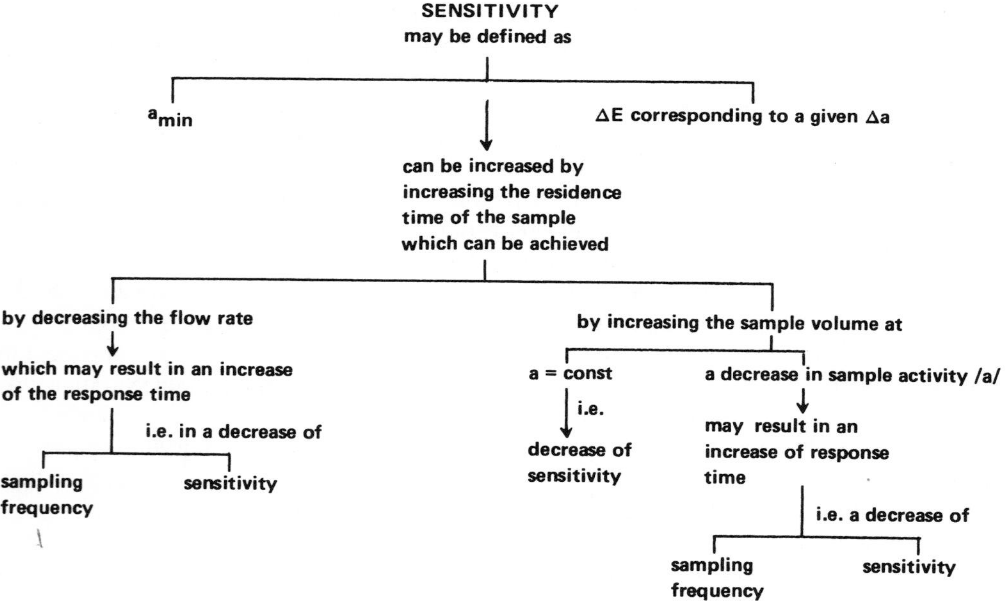

FIGURE 2. Schematic illustration of the different possibilities for increasing the sensitivity in flow injection analysis.

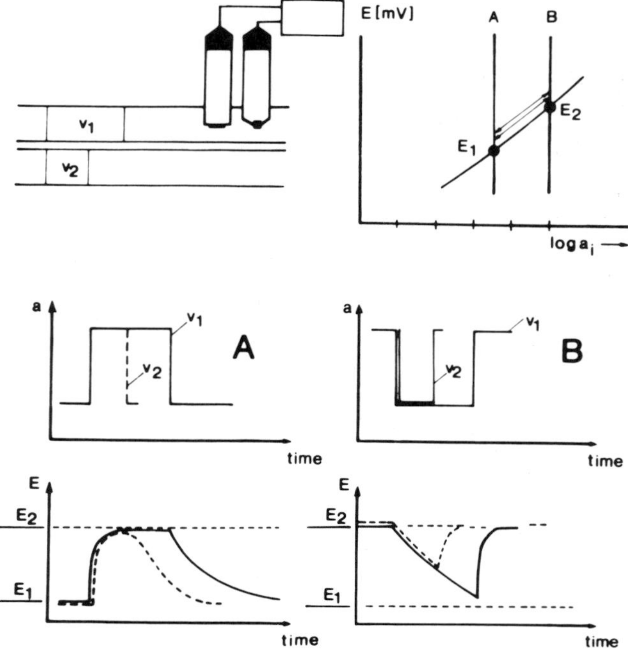

FIGURE 3. Schematic representation of the input and output signals of a potentiometric cell for two different injected sample volumes in flow injection analysis. In case A, the concentration of the ion to be determined in the flowing background electrolyte is smaller than that of the injected sample. In case B, the opposite is true.

solution with a higher primary ion concentration compared to that of the sample injected may also be advantageous as the baseline variation is concerned. Higher background primary ion concentration level results in a better reproducibility.

In the above discussion, it was intended to emphasize that in selecting the appropriate residence time in a flow injection system, a judicious choice must be made between the disadvantages with respect to response time produced by decreasing flow rate, and the problems related to both sensitivity and response time which may be faced with large sample dispersion.

The evaluation of the recorded signals is more complex or the determination of the concentration of the injected sample may even by impossible if the latter contains interfering ions in varying concentrations.[254] Two potential peaks of opposite direction can generally be recorded (Figure 4) when the injected samples contain interfering ions corresponding to the "two-ion range". Under these conditions, by decreasing the injected volume, the time interval corresponding to the first peak will only be decreased (Figure 4). Thus, in complete agreement with the theoretical expectations relevant to the dynamic response characteristics of the sensors in the two-ion range, asymmetric potential peaks are often reported in the literature by injecting samples of high interfering ion activity.[254]

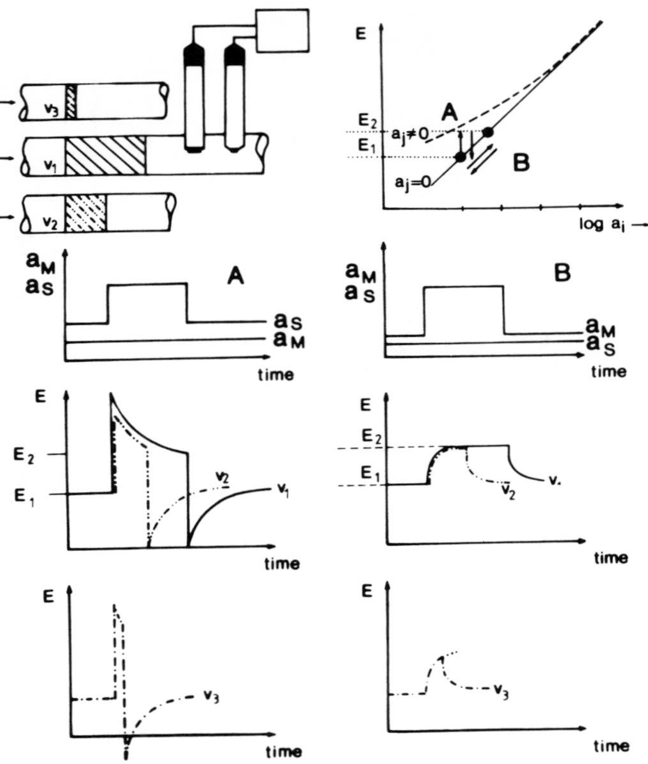

FIGURE 4. Schematic representation of the input and output signals of a potentiometric cell for three different injected volumes in flow injection analysis if samples containing interfering ions are injected. In case A, the primary ion activity of the sample is the same as that of the streaming solution and is kept constant. In case B, the interfering ion activity is kept constant and that of the primary ion is altered.

REFERENCES

126

1. **MacDonald, D. D.,** *Transient Techniques in Electrochemistry,* Plenum Press, Nev
2. **Vetter, K. J.,** *Electrochemical Kinetics, Theoretical Aspects,* Academic Press, Ne
3. **Bard, A. J. and Faulkner, L. R.,** *Electrochemical Methods, Fundamental Appli* Sons, New York, 1980.
4. **Tóth, K., Gavallér, I., and Pungor, E.,** *Anal. Chim. Acta,* 57, 131, 1971.
5. **Tóth, K. and Pungor, E.,** *Anal. Chim. Acta,* 64, 417, 1973.
6. **Lindner, E., Tóth, K., and Pungor, E.,** *Anal. Chem.,* 48, 1071, 1976.
7. **Morf, W. E., Lindner, E., and Simon, W.,** *Anal. Chem.,* 47, 1596, 1975.
8. **Lindner, E.,** Dynamic Response of Ion-Selective Electrodes, Dissertation, Cand. Chem. Sci., Budapest, 1983.
9. **Buck, R. P.,** *Ion Sel. Electrode Rev.,* 4, 3, 1982.
10. **Buck, R. P.,** in *Ion-Selective Electrodes in Analytical Chemistry,* Freiser, H., Ed., Plenum Press, New York, 1978, chap. 1.
11. **Buck, R. P.,** *CRC Crit. Rev. Anal. Chem.,* 5, 323, 1975.
12. **Buck, R. P.,** *Hung. Sci. Instrum.,* 49, 7, 1980.
13. **Cammann, K.,** in *Ion-Selective Electrodes,* Pungor, E., Ed., Akadémiai Kiadó, Budapest, 1978, 297.
14. **Cammann, K. and Rechnitz, G. A.,** in *Ion-Selective Electrodes,* Vol. 4, Pungor, E., Ed., Akadémiai Kiadó, Budapest, 1985, 35.
15. **Morf, W. E., Wuhrmann, P., and Simon, W.,** *Anal. Chem.,* 48, 1031, 1976.
16. **Lindner, E., Wuhrmann, P., Simon, W., and Pungor, E.,** in *Ion-Selective Electrodes,* Pungor, E., Ed., Akadémiai Kiadó, Budapest, 1977, 159.
17. **Morf, W. E.,** The principles of ion-selective electrodes and of membrane transport, in *Studies in Analytical Chemistry 2,* Pungor, E., Simon, W., and Inczédy, J., Eds., Akadémiai Kiadó, Budapest, 1981.
18. **Buck, R. P.,** *Anal. Chem.,* 44, 270R, 1972.
19. **Buck, R. P.,** *Anal. Chem.,* 46, 28R, 1974.
20. **Buck, R. P.,** *Anal. Chem.,* 48, 23R, 1976.
21. **Buck, R. P.,** *Anal. Chem.,* 50, 17R, 1978.
22. **Buck, R. P.,** in *Ion-Selective Electrodes,* Pungor, E., Ed., Akadémiai Kiadó, Budapest, 1978, 21.
23. **Stover, F. S., Brumleve, T. R., and Buck, R. P.,** *Anal. Chim. Acta,* 109, 259, 1979.
24. **Buffle, J. and Parthasarathy, N.,** *Anal. Chim. Acta,* 93, 111, 1977.
25. **Parthasarathy, N., Buffle, J., and Haerdi, N.,** *Anal. Chim. Acta,* 93, 121, 1977.
26. **Sluyters-Rehbach, M. and Sluyters, J. H.,** in *Electroanalytical Chemistry,* Vol. 4, Bard, A. J., Ed., Marcel Dekker, New York, 1969, Chap. 1.
27. **Macdonald, J. R.,** Interpretation of AC Impedance Measurements in Solids, in *Superionic Conductors,* Maha, G. D. and Roth, W. L., Eds., Plenum Press, New York, 1976, 81.
28. **Macdonald, J. R.,** *J. Electroanal. Chem. Interfacial Electrochem.,* 53, 1, 1974.
29. **Archer, W. I. and Armstrong, R. D.,** *The Application of A.C. Impedance Methods to Solid Electrolytes,* Spec. Period. Rep., The Chemical Society of London, 1980, 157.
30. **Sandifer, J. R. and Buck, R. P.,** *J. Electroanal. Chem. Interfacial Electrochem.,* 56, 385, 1974.
31. **Warburg, E.,** *Ann. Phys., Neve Folge,* 67, 493, 1899.
32. **Warburg, E.,** *Ann. Phys., Vierte Folge,* 6, 125, 1901.
33. **Eisenman, G.,** *Glass Electrodes for Hydrogen and Other Cations: Principles and Practice,* Marcel Dekker, New York, 1967.
34. **Cammann, K.,** *Working with Ion-Selective Electrodes,* Springer-Verlag, New York, 1979.
35. **Vlasov, Yu. G.,** *Ion-Selective Electrodes,* Vol. 3, Pungor, E., Ed., Akadémiai Kiadó, Budapest, 1981, 147.
36. **Ammann, D., Pretsch, E., Simon, W., Lindner, E., Bezegh, A., and Pungor, E.,** *Anal. Chim. Acta,* 171, 119, 1985.
37. **Pretsch, E., Wegmann, D., Ammann, D., Bezegh, A., Dinten, O., Läubli, M. W., Morf, W. E., Oesch, U., Sugahara, K., Weiss, H., and Simon, W.,** in *Recent Advances in the Theory and Application of Ion-Selective Electrodes in Physiology and Medicine,* Kessler, M., Harrison, D. K., and Höpler, J., Eds., Springer-Verlag, Basel, 1985.
38. **Nieman, T. A. and Horvai, G.,** *Anal. Chim. Acta,* 170, 359, 1985.
39. **Horvai, G., Nieman, T. A., and Pungor, E.,** in *Ion-Selective Electrodes,* Vol. 4, Pungor, E., Ed., Akadémiai Kiadó, Budapest, 1985, 439.
40. **Macdonald, J. R.,** *J. Chem. Phys.,* 61, 3977, 1974.
41. **Franceschetti, D. R. and Macdonald, J. R.,** *J. Electroanal Chem. Interfacial Electrochem.,* 82, 271, 1977.
42. **MacDonald, J. R.,** *J. Electroanal. Chem. Interfacial Electrochem.,* 70, 17, 1976.

43. **MacDonald, J. R. and Franceschetti, D. R.,** *J. Phys. Chem.,* 68, 1614, 1978.
44. **Buck, R. P.,** *J. Electroanal. Chem. Interfacial Electrochem.,* 18, 381, A68.
45. **Buck, R. P., Mathis, D. E., and Rhodes, R. K.,** *J. Electroanal. Chem. Interfacial Electrochem.,* 80, 245, 1977.
46. **Buck, R. P.,** *J. Electroanal. Chem. Interfacial Electrochem.,* 18, 363, 1968.
47. **Buck, R. P. and Krull, J.,** *J. Electroanal. Chem. Interfacial Electrochem,.,* 18, 387, 1968.
48. **Brand, M. J. D. and Rechnitz, G. A.,** *Anal. Chem.,* 41, 1788, 1969.
49. **Rhodes, R. K. and Buck, R. P.,** *Anal. Chim. Acta,* 113, 67, 1980.
50. **Brand, M. J. D. and Rechnitz, G. A.,** *Anal. Chem.,* 42, 476, 1970.
51. **Rhodes, R. K. and Buck, R. P.,** *J. Electroanal. Chem. Interfacial Electrochem.,* 86, 349, 1978.
52. **Rhodes, R. K. and Buck, R. P.,** *Anal. Chim. Acta,* 110, 185, 1979.
53. **Mertens, J. and Van den Winkel, P.,** Vereecken, J., *Biochem. Bioenerg.,* 5, 699, 1978.
54. **Mathis, D. E. and Buck, R. P.,** *J. Membr. Sci.,* 4, 379, 1979.
55. **Mathis, D. E., Stover, F. S., and Buck, R. P.,** *J. Membr. Sci.,* 4, 395, 1979.
56. **Armstrong, R. D., Covington, A. K., Evans, G. P., and Handyside, T.,** *Electrochim. Acta,* 29, 1127, 1984.
57. **Armstrong, R. D., Covington, A. K., and Evans, G. P.,** *J. Electroanal. Chem. Interfacial Electrochem.,* 159, 33, 1983.
58. **Ahmad-Bitar, R., Abdul-Gader, M. M., Zihlif, A. M., and Jaber, A. M. Y.,** *J. Electroanal. Chem. Interfacial Electrochem.,* 143, 121, 1983.
59. **Horvai, G., Gráf, E., Tóth, K., Pungor, E., and Buck, R. P.,** *Anal. Chem.,* 58, 2735, 1986.
60. **Tóth, K., Gráf, E., Horvai, G., Pungor, E., and Buck, R. P.,** *Anal. Chem.,* 58, 2741, 1986.
61. **Buck, R. P., Tóth, K., Gráf, E., and Pungor, E.,** Role of plasticized poly(vinyl chloride) properties on characteristics of valinomycin neutral ion-selective electrodes. III. Donnan exclusion failure in low anion site density membranes, *J. Electroanal. Chem. Interfacial Electrochem.,* in preparation.
62. **Koebel, M.,** Zum Austauschverhalten des Aquotisierten Silberions an Silber und Silberionen-Festelektrolyten. Die gemischte Leitung von Silbersulfid bei Raumtemperatur, Dissertation, ETH, Zürich, 1972.
63. **Koebel, M., Ibl, N., and Frei, A. M.,** *Electrochim. Acta,* 19, 287, 1974.
64. **Cammann, K.,** Untersuchungen zur Wirkungsweise Ionenselektiven Elektroden (Abschätzung von Standard-austauschstromdichten), Dissertation, Ludwig-Maximilians-Universität, München, 1975.
65. **Cammann, K. and Rechnitz, G. A.,** *Anal. Chem.,* 48, 856, 1976.
66. **Cammann, K.,** *Anal. Chem.,* 50, 936, 1978.
67. **Cammann, K.,** *Instrum. Forsch.,* 9, 1, 1982.
68. **Buttler, J. A. V.,** *Trans. Faraday Soc.,* 19, 734, 1924.
69. **Erdey-Gruz, T. and Volmer, M.,** *Z. Phys. Chem.,* 150A, 203, 1930.
70. **Meier, P. C., Morf, W. E., Läubli, M. W., and Simon, W.,** *Anal. Chim. Acta,* 156, 1, 1984.
71. **Morf, W. E., Ammann, D., and Simon, W.,** *Chimia,* 28, 65, 1974.
72. **Morf, W. E., Kahr, G., and Simon, W.,** *Anal. Lett.,* 7(1), 9, 1974.
73. **Sand, H. J. S.,** *Z. Phys. Chem.,* 35, 641, 1900.
74. **Berzins, T. and Delahay, P.,** *J. Am. Chem. Soc.,* 77, 6448, 1955.
75. **Berzins, T. and Delahay, P.,** *Z. Electrochem.,* 59, 792, 1955.
76. **Delahay, P. and Berzins, T.,** *J. Am. Chem. Soc.,* 75, 2486, 1953.
77. **Donnan, F. G.,** *Z. Electrochem.,* 17, 572, 1911.
78. **Donnan, F. G. and Guggenheim, E. A.,** *Z. Phys. Chem.,* A162, 346, 1932.
79. **Theorell, T.,** *Proc. Soc. Exp. Biol. Med.,* 33, 282, 1935.
80. **Meyer, K. H. and Sievers, J. F.,** *Helv. Chim. Acta,* 19, 649, 1936; 19, 665, 1936; 19, 987, 1936.
81. **Helferich, F.,** *Ionenaustauschen, Grundlagen,* Vol. 1, Verlag Chemie, Weinheim/Bergstrasse, 1959.
82. **Schlögl, R.,** *Z. Phys. Chem. (Frankfurt am Main),* 1, 305, 1954.
83. **Schlögl, R.,** *Stofftransport durch Membranen,* Steinkopff, Darmstadt, 1964.
84. **Eisenman, G., Rudin, D. O., and Casby, J. U.,** *Science,* 126, 871, 1957.
85. **Conti, F. and Eisenman, G.,** *Biophys. J.,* 5, 247, 1965; 5, 211, 1965.
86. **Eisenman, G.,** in *Ion-Selective Electrodes,* Durst, R. A., Ed., N.B.S. Spec. Publ. No. 314, Washington, D.C., 1969.
87. **Wuhrmann, H. R., Morf, W. E., and Simon, W.,** *Helv. Chim. Acta,* 56, 1011, 1973.
88. **Nikolsky, B. P.,** *Zh. Fiz. Khim.,* 10, 495, 1937.
89. **Nikolsky, B. P., Shultz, M. M., Belijustin, A. A., and Lev, A. A.,** in *Glass Electrodes for Hydrogen and Other Cations: Principles and Practice,* Eisenman, G., Ed., Marcel Dekker, New York, 1967, chap. 6.
90. **Pungor, E.,** *Anal. Chem.,* 39, 28A, 1967.
91. **Pungor, E. and Tóth, K.,** *Analyst,* 95, 625, 1970.
92. **Buck, R. P.,** *Anal. Chem.,* 40, 1432, 1968.

93. **Cammann, K.**, Ion-selective bulk membranes as models for biomembranes, in *Membranes*, Springer-Verlag, New York, in press.
94. **Iljuschemko, M. A. and Mirkin, W. A.**, *Chabarschysny Vestnik*, Kasakstan Academy of Science, 4, 41, 1981 (translated into English in Reference 93).
95. **Koryta, J.**, *Ion Sel. Electrode Rev.*, 5, 131, 1983.
96. **Oesch, U. and Simon, W.**, *Helv. Chim. Acta*, 62, 754, 1979.
97. **Oesch, U. and Simon, W.**, *Anal. Chem.*, 52, 692, 1980.
98. **Yoshida, S. and Hayano, S.**, *J. Membr. Sci.*, 11, 157, 1982.
99. **Chaudhari, S. N. K. and Cheng, K. L.**, *Mikrochim. Acta (Wien)*, 2, 159, 1980.
100. **Harsányi, E. G., Tóth, K., Pólos, L., and Pungor, E.**, *Anal. Chem.*, 54, 1094, 1982.
101. **Harsányi, E. G., Tóth, K., and Pungor, E.**, *Anal. Chim. Acta*, 152, 163, 1983.
102. **Harsányi, E. G., Tóth, K., and Pungor, E.**, in *Ion-Selective Electrodes*, Vol. 4 Akadémiai Kiadó, Budapest, 1985, 399.
103. **Lindner, E.**, unpublished results.
104. **Tóth, K.**, Ion-Selective Membrane Electrodes, Dissertation, Cand. Chem. Sci., Veszprém, Hungary, 1969.
105. **Thoma, A. P., Viviani-Naurer, A., Arvanitis, S., Morf, W. E., and Simon, W.**, *Anal. Chem.*, 49, 1597, 1977.
106. **Zimens, K. E.**, *Ark. Kemi Mineral. Geol.*, 16, 23A, 1946.
107. **Reber, K.**, *Phot. Korr*, 105, 175, 1969; 105, 191, 1969.
108. **Kirsch, N. N. L. and Simon, W.**, *Helv. Chim. Acta*, 59, 357, 1976.
109. **Kirsch, N. N. L.**, Die Bedeutung von Komplexbildungs — und Extraktionsgleichgewichten für die Alkali — und Erdalkali Metallionenselektivität von Flüssigmembran, Elektroden Beruhend auf Azyklischen, Ungeladenen Liganden, Dissertation, ETH, Zürich, 1976.
110. **Wuhrmann, P., Thoma, A. P., and Simon, W.**, *Chimia*, 27, 637, 1973.
111. **Kellner, R., Fischböck, G., Götzinger, G., Pungor, E., Tóth, K., Pólos, L., and Lindner, E.**, *Fresenius Z. Anal. Chem.*, 322, 151, 1985.
112. **Kellner, R., Götzinger, G., Pungor, E., Tóth, K., and Pólos, L.**, *Fresenius Z. Anal. Chem.*, 319, 839, 1984.
113. **Peschanski, Mme. D.**, *J. Chim. Phys.*, 47, 933, 1950.
114. **Leo, A., Hansch, D., and Elkins, D.**, *Chem. Rev.*, 71, 525, 1971.
115. **Disteche, A. and Dubuisson, M.**, *Rev. Sci. Instrum.*, 25, 869, 1954.
116. **Efstathiou, C. E., Konpparis, M. A., and Hadjiioannou, T. P.**, *Ion Sel. Electrode Rev.*, 7, 203, 1985.
117. **Meier, J. and Schwarzenbach, G.**, *Helv. Chim. Acta*, 40, 907, 1957.
118. **Sirs, J. A.**, *Trans. Faraday Soc.*, 54, 207, 1958.
119. **Chaplin, A. L.**, Application of Industrial pH Control, Pittsburg, 1950.
120. **Lengyel, I. and Solti, M.**, pH-Mérés és *Szabályozás, Müszaki Könyvkiadó*, Budapest, 1967.
121. **Mattock, G.**, *Analyst*, 87, 930, 1962.
122. **Rechnitz, G. A.**, *Talanta*, 11, 1467, 1964.
123. **Savage, J. A. and Isard, J. O.**, *J. Phys. Chem. Glasses*, 3, 147, 1962.
124. **Beck, W. H., Caudle, J., Covington, A. K., and Wynne-Jones, W. F. K.**, *Proc. Chem. Soc. London*, p. 110, 1963.
125. **Rechnitz, G. A., Kresz, M. R., and Zamochnik, S. B.**, *Anal. Chem.*, 38, 973, 1966.
126. **Rechnitz, G. A. and Kresz, M. R.**, *Anal. Chem.*, 38, 1786, 1966.
127. **Hseu, T. M. and Rechnitz, G. A.**, *Anal. Chem.*, 40, 1054, 1968.
128. **Lindner, E., Tóth, K., and Pungor, E.**, *Pure Appl. Chem.*, 58, 469, 1986.
129. **Oehme, M. and Simon, W.**, *Anal. Chim. Acta*, 86, 21, 1976.
130. **Morf, W. E., Oehme, M., and Simon, W.**, in *Ionic Actions on Vascular Smooth Muscle*, Betz, E., Ed., Springer-Verlag, New York, 1976.
131. **Morf, W. E. and Simon, W.**, *Hung. Sci. Instrum.*, 41, 1, 1977.
132. **Morf, W. E. and Simon, W.**, in *Ion-Selective Electrodes in Analytical Chemistry*, Vol. 1, Freiser, H., Ed., Plenum Press, New York, 1978.
133. **Light, T. S. and Schwartz, J. L.**, *Anal. Lett.*, 1, 825, 1968.
134. **Shatkay, A.**, *Anal. Chem.*, 48, 1039, 1976.
135. **Karlberg, B.**, *J. Electroanal. Chem. Interfacial Electrochem.*, 45, 127, 1973.
136. **Hawkings, R. C., Corriveau, L. P. V., Kushneriuk, S. A., and Wong, P. W.**, *Anal. Chim. Acta*, 102, 61, 1978.
137. **Karlberg, B.**, *J. Electroanal. Chem. Interfacial Electrochem.*, 42, 115, 1973.
138. **Karlberg, B.**, *J. Electroanal. Chem. Interfacial Electrochem.*, 49, 1, 1974.
139. **Fleet, B., Ryan, T. H., and Brand, M. J. D.**, *Anal. Chem.*, 46, 12, 1974.
140. **Johansson, G. and Norberg, K.**, *J. Electroanal. Chem. Interfacial Electrochem.*, 18, 239, 1968.
141. **Mertens, J., Van den Winkel, P., and Massart, D. L.**, *Anal. Chem.*, 48, 272, 1976.

142. **Markovic, P. L. and Osburn, J. O.,** *AICHE J.*, 19, 504, 1973.
143. **Vandeputte, M., Dryon, L., and Massart, D. L.,** in *Ion-Selective Electrodes,* Pungor, E., Ed., Akadémiai Kiadó, Budapest, 1978, 583.
144. **Rechnitz, G. A. and Kugler, G. C.,** *Anal. Chem.*, 39, 1682, 1967.
145. **Rangarajan, R. and Rechnitz, G. A.,** *Anal. Chem.*, 47, 324, 1975.
146. **Denks, A. and Neeb, R.,** *Fresenius Z. Anal. Chem.*, 285, 233, 1977.
147. **Degawa, H., Shinozuka, N., and Hayano, S.,** *Chem. Lett.*, p. 25, 1983.
148. **Nagy, K. and Fjeldly, T. A.,** in *Ion-Selective Electrodes,* Vol. 3, Pungor, E., Ed., Akadémiai Kiadó, Budapest, 1981, 287.
149. **Lindner, E., Tóth, K., Berube, T. R., and Pungor, E.,** Wissentschaftliche Beiträge der Karl Marx Universität, Leipzig, 1986, 67; *Anal. Chem.*, in press.
150. **Hill, A. V.,** *Proc. R. Soc. London Ser. B,* 135, 446, 1948.
151. **Lindner, E., Tóth, K., and Pungor, E.,** *Bunseki Kagaku,* 30, S67, 1981.
152. **Lindner, E., Tóth K., and Pungor, E.,** *Anal. Chem.*, 54, 72, 1982.
153. **Lindner, E., Tóth, K., and Pungor, E.,** *Magy. Kém. Foly.*, 88, 49, 1982.
154. **Tóth, K.,** in *Ion-Selective Electrodes,* Pungor, E., Ed., Akadémiai Kiadó, Budapest, 1973, 145.
155. **Ammann, D., Pretsch, E., and Simon, W.,** *Anal. Lett.*, 7, 23, 1974.
156. **Bladel, J. W. and Dinwiddie, D. E.,** *Anal. Chem.*, 47, 1070, 1975.
157. **Blaedel, J. W. and Dinwiddie, D. E.,** *Anal. Chem.*, 46, 873, 1974.
158. **Bock, R. and Puff, H. J.,** *Fresenius Z. Anal. Chem.*, 240, 381, 1968.
159. **Bock, R. and Stecker, S.,** *Fresenius Z. Anal. Chem.*, 235, 322, 1968.
160. **Denks, A. and Neeb, R.,** *Fresenius Z. Anal. Chem.*, 297, 121, 1979.
161. **Denks, A.,** Vergleichende Untersuchungen über das Ansprechverhalten von metalischen Ag-Elektroden und $Ag_2S/AgX(x = Cl,I)$. Festkörpermembranelektroden bei schnellen Konzentrationsänderungen in Strömenden lösungen Dissertation, Johannes Gutenberg-Universität, Mainz, 1977.
162. **Orion Research Inc.,** *Newsletter,* 3(1 and 2), 8, 1971.
163. **Uemasu, I. and Umezawa, Y.,** *Anal. Chem.*, 54, 1198, 1982.
164. **Umezawa, Y., Tasaki, I., and Fujiwara, S.,** in *Ion-Selective Electrodes,* Vol. 3, Pungor, E., Ed., Akadémiai Kiadó, Budapest, 1981, 359.
165. **Lindner, E., Tóth, K., Pungor, E., and Umezawa, Y.,** *Anal. Chem.*, 56, 810, 1984.
166. **Alexander, P. W. and Rechnitz, G. A.,** *Anal. Chem.*, 46, 250, 1974.
167. **Buck, R. P. and Boles, J. H., Porter, R. D., and Margolis, J. A.,** *Anal. Chem.*, 46, 255, 1974.
168. **Karlberg, B.,** *Anal. Chim. Acta,* 66, 93, 1973.
169. **Decker, U. P. and Brott, R.,** *Z. Chem.*, 24, 77, 1984.
170. **Decker, U. P. and Beckhaus, S.,** *Z. Chem.*, 25, 417, 1985.
171. **Harzdorf, C. and Henning, G.,** in *Ion-Selective Electrodes,* Pungor, E., Ed., Akadémiai Kiadó, Budapest, 1978, 379.
172. **Sorentino, M. H. and Rechnitz, G. A.,** *Anal. Chem.*, 46, 943, 1974.
173. **Rechnitz, G. A. and Lin, Z. F.,** *Anal. Chem.*, 39, 1406, 1967.
174. **Morf, W. E.,** *Anal. Lett.*, 10, 87, 1977.
175. **Bagg, J. and Vinen, R.,** *Anal. Chem.*, 44, 1773, 1972.
176. **Reinsfelder, R. E. and Schultz, F. A.,** *Anal. Chim. Acta,* 65, 425, 1973.
177. **Tomita, T.,** in *Glass Microelectrodes,* Lavalle, M., Schane, O. F., and Herbert, N. C., Eds., John Wiley & Sons, New York, 1969, chap. 10.
178. **Grundfest, H.,** in *Glass Microelectrodes,* Lavalle, M., Schane, O. F., and Herbert, N. C., Eds., John Wiley & Sons, New York, 1969, chap. 8.
179. **Rechnitz, G. A.,** in *Ion-Selective Electrodes,* Durst, R. A., Ed., NBS Spec. Publ. No. 314, Washington, D.C., 1969, chap. 10.
180. **Lindner, E., Tóth, K., and Pungor, E.,** *Anal. Chem.*, 54, 202, 1982.
181. **Lindner, E., Tóth, K., Pungor, E., and Nowakowski, K.,** *Magy. Kém. Foly.*, 88, 55, 1982.
182. **Tóth, K., Lindner, E., and Pungor, E.,** in *Ion-Selective Electrodes,* Vol. 3, Akadémiai Kiadó, Budapest, 1981, 135.
183. **Morf, W. E.,** *Anal. Chem.*, 55, 1165, 1983.
184. **Gratzl, M., Lindner, E., and Pungor, E.,** *Anal. Chem.*, 57, 1506, 1985.
185. **Gratzl, M., Lindner, E., and Pungor, E.,** *Magy. Kém. Foly.*, 91, 101, 1985.
186. **Lindner, E., Gratzl, M., and Pungor, E.,** in *Ion-Selective Electrodes,* Vol. 4, Pungor, E., Ed., Akadémiai Kiadó, Budapest, 1985, 179.
187. **Kennedy, C. D.,** *Analyst,* 108, 1003, 1983.
188. **Beck, W. H. and Wynne-Jones, W. F. K.,** *J. Chim. Phys. Phys. Chim. Biol.*, 49, C97, 1952.
189. **Perley, G. A.,** *Anal. Chem.*, 21, 559, 1949.
190. **Johansson, G., Karlberg, B., and Wikby, A.,** *Talanta,* 22, 953, 1975.

191. **Jaenicke, W.,** *Z. Electrochem.,* 55, 648, 1951.
192. **Jaenicke, W.,** *Z. Electrochem.,* 57, 843, 1953.
193. **Jaenicke, W., and Haase, M.,** *Z. Electrochem.,* 63, 521, 1959.
194. **Van den Winkel, P., Mertens, J., and Massart, D. L.,** *Anal. Chem.,* 46, 1765, 1974.
195. **Liberti, A.,** in *Ion-Selective Electrodes,* Pungor, E., Ed., Akadémiai Kiadó, Budapest, 1973, 37.
196. **Kahr, G.,** Beitrag zum Elektromotorischen Verhalten von Ionenselektiven Festkörpermembranelektroden, Dissertation, ETH, Zürich, 1972.
197. **Malissa, H., Grasserbauer, M., Pungor, E., Tóth, K., Pápay, M. K., and Pólos, L.,** *Anal. Chim. Acta,* 80, 223, 1975.
198. **Pungor, E. Tóth, K., Pápay, M. K., Pólos, L., Malissa, H., Grasserbauer, M., Hoke, E., Ebel, M. F., and Persey, K.,** *Anal. Chim. Acta,* 109, 279, 1979.
199. **Pungor, E., Gratzl, M., Pólos, L., Tóth, K., Ebel, M. F., Ebel, H., Zuba, G., and Wernish, J.,** *Anal. Chim. Acta,* 156, 9, 1984.
200. **Ebel, M. F., Tóth, K., Pólos, L., and Pungor, E.,** *Surf. Interface Anal.,* 2(5), 197, 1970.
201. **Bauke, F. G. K.,** in *Ion-Selective Electrodes,* Pungor, E., Ed., Akadémiai Kiadó, Budapest, 1977, 215.
202. **Bouquet, G., Dobos, S., and Boksay, Z.,** *Ann. Univ. Sci. Budapest Sect. Chim.,* 6, 5, A64.
203. **Boksay, Z., Bouquet, G., and Dobos, S.,** *Phys. Chem. Glasses,* 8, 140, 1967.
204. **Gulens, J.,** *Ion Sel. Electrode Rev.,* 2, 117, 1981.
205. **Gratzl, M., Gryzelkó, L., Kömives, J., Tóth, K., and Pungor, E.,** in *Ion Selective Electrodes,* Vol. 4, Pungor, E., Ed., Akadémiai Kiadó, Budapest, 1985, 417.
206. **Rechnitz, G. A. and Hameka, H. F.,** *Fresenius Z. Anal. Chem.,* 214, 252, 1965.
207. **Morf, W. E., Kahr, G., and Simon, W.,** *Anal. Chem.,* 46, 1538, 1974.
208. **Bauke, F. G. K.,** *Electrochim. Acta,* 17, 851, 1972.
209. **Hulanicki, A. and Lewenstam, A.,** *Anal. Chem.,* 53, 1401, 1981.
210. **Hulanicki, A. and Lewenstam, A.,** *Talanta,* 24, 171, 1977.
211. **Nernst, W.,** *Z. Phys. Chem.,* 47, 52, 1904.
212. **Evans, D. H.,** *Anal. Chem.,* 44, 875, 1972.
213. **Bound, G. P., Fleet, B., von Storp, H., and Evans, D. H.,** *Anal. Chem.,* 45, 788, 1973.
214. **Carslaw, H. S. and Jaeger, J. C.,** *Conduction of Heat in Solids,* Oxford University Press, London, 1959.
215. **Cranck, J.,** *The Mathematics of Diffusion,* Oxford University Press, London, 1956.
216. **Ross, J. W., Riseman, J. H., and Krueger, J. A.,** *Pure Appl. Chem.,* 36, 473, 1973.
217. **Bailey, P. L. and Riley, M.,** *Analyst (London),* 100, 145, 1975.
218. **Arnold, M. A. and Meyerhoff, M. E.,** *Anal. Chem.,* 56, 20R, 1984.
219. **Morf, W. E., Mostert, I. A., and Simon, W.,** *Anal. Chem.,* 57, 1122, 1985.
220. **Arnold, M. A. and Rechnitz, G. A.,** *Anal. Chim. Acta,* 158, 379, 1984.
221. **Van der Schoot, B. and Bergveld, P.,** *Anal. Chim. Acta,* 166, 93, 1984.
222. **Lindner, E., Tóth, K., Pungor, E., Morf, W. E., and Simon, W.,** *Anal. Chem.,* 50, 1627, 1978.
223. **Vielstich, W.,** *Z. Electrochem.,* 57, 646, 1953.
224. **Levich, V. G.,** *Physicochemical Hydrodynamics,* Prentice Hall, Englewood Cliffs, N.J., 1962.
225. **Davies, J. T.,** *Turbulence Phenomena,* Academic Press, New York, 1972.
226. **Denks, A. and Neeb, R.,** *Fresenius Z. Anal. Chem.,* 298, 131, 1979.
227. **Kortüm, G. and Bockris, J. O. M.,** *Textbook of Electrochemistry,* Elsevier, New York, 1951.
228. **Erdey-Gruz, T. and Shay, G.,** *Elméleti Fizikai Kémia,* Vol. 2, Tankönyvkiadó, Budapest, 1964, 474.
229. **Frost, A. A. and Pearson, R. G.,** *Kinetics and Mechanism: A Study of Homogeneous Chemical Reactions,* John Wiley & Sons, New York, 1958.
230. **Müller, R. H.,** *Anal. Chem.,* 41, 113A, 1969.
231. **Davies, C. W. and Jones, A. L.,** *Trans. Faraday Soc.,* 51, 812, 1955.
232. **Nancollas, G. H.,** *Croat. Chim. Acta,* 45, 225, 1973.
233. **Štefanac, Z. and Simon, W.,** *Chimia (Switzerland),* 20, 436, 1966.
234. **Štefanac, Z. and Simon, W.,** *Microchem. J.,* 12, 125, 1967.
235. **Pioda, L. A. R., Stankova, V., and Simon, W.,** *Anal. Lett.,* 2, 665, 1969.
236. **Fiedler, U. and Růžička, J.,** *Anal. Chim. Acta,* 67, 179, 1973.
237. **Pick, J., Tóth, K., Pungor, E., Vasak, M., and Simon, W.,** *Anal. Chim. Acta,* 64, 477, 1973.
238. **Boles, J. H. and Buck, R. P.,** *Anal. Chem.,* 45, 2057, 1973.
239. **Kedem, O., Perry, M., and Bloch, R.,** IUPAC Int. Symp. Selective Ion-Sensitive Electrodes, Paper No. 44, University of Wales, Cardiff, 1973.
240. **Ammann, D., Morf, W. E., Anker, P., Meier, P. C., Pretsch, E., and Simon, W.,** *Ion Sel. Electrode Rev.,* 5, 3, 1983.
241. **Shatkay, A. and Azor, M.,** *Anal. Chim. Acta,* 133, 183, 1981.
242. **Belijustin, A. A., Valova, I. V., and Ivanovskaja, I. S.,** in *Ion-Selective Electrodes,* Pungor, E., Ed., Akadémiai Kiadó, Budapest, 1978, 235.

243. **Belijustin, A. A.,** *Usp. Khim.*, 49, 1880, 1980.
244. **Tóth, K., Nagy, G., and Pungor, E.,** in *Ion-Selective Electrode* Methodology, Vol. 2, Covington, A. K., Ed., CRC Press, Boca Raton, Fla., 1979, chap. 4.
245. **Pungor, E., Fehér, Zs., and Váradi, M.,** *Crit. Rev. Anal. Chem.*, 9, 97, 1980.
246. **Nagy, G., Fehér, Zs., Tóth, K., and Pungor, E.,** *Hung. Sci. Instrum.*, 41, 27, 1977.
247. **Tóth, K., Nagy, G., Fehér, Zs., Horvai, G., and Pungor, E.,** *Anal. Chim. Acta,* 114, 48, 1980.
248. **Tóth, K., Fucskó, J., Lindner, E., Fehér, Zs., and Pungor, E.,** *Anal. Chim. Acta,* 179, 359, 1986.
249. **Shatkay, A. and Hayano, S.,** *Anal. Chem.*, 57, 366, 1985.
250. **Schwab, G. M.,** *Kolloid Z.*, 101, 204, 1942.
251. **Senkyr, J. and Petr, J.,** in *Ion-Selective Electrodes,* Vol. 3, Pungor, E., Ed., Akadémiai Kiadó, Budapest, 1981, 327.
252. **Morf, W. E.,** in *Ion-Selective Electrodes,* Vol. 3, Pungor, E., Ed., Akadémiai Kiadó, Budapest, 1981, 267.
253. **Jyo, A. and Ishibashi, N.,** in *Studies in Analytical Chemistry,* Vol. 2, Pungor, E., Simon, W., and Inczédy, J., Eds., Akadémiai Kiadó, Budapest, 1981, 246.
254. **Hansen, E. H., Ghose, A. K., and Růžička, J.,** *Analyst,* 102, 705, 1977.
255. **Moody, G. J. and Thomas, J. D. R.,** *Selective Ion-Sensitive Electrodes,* Merrow, Watford, England, 1970.
256. **Doremus, R. H.,** in *Glass Electrodes for Hydrogen and Other Cations: Principles and Practice,* Marcel Dekker, New York, 1967, chap. 4.
257. **Brand, M. J. D., and Rechnitz, G. A.,** *Anal. Chem.*, 41, 1185, 1969.
258. **Vesely, J. and Stulik, K.,** *Anal. Chim. Acta,* 73, 157, 1974.
259. **Hulanicki, A.,** in *Ion-Selective Electrodes,* Vol. 3, Pungor, E., Ed., Akadémiai Kiadó, Budapest, 1981, 103.
260. **Robinson, R. A. and Stokes, R. H.,** *Electrolyte Solutions,* Butterworths, London, 1955.
261. **Hulanicki, A., Lewenstam, A., and Maj-Zurawska, M.,** *Anal. Chim. Acta,* 107, 21, 1979.
262. **Pungor, E., Tóth, K., and Hrabéczy-Páll, A.,** *Pure Appl. Chem.*, 51, 1913, 1979.
263. **Guilbault, G. G.,** *IUPAC Int. Bull. No. 1.*, 69, 1978.
264. **IUPAC,** *Pure Appl. Chem.*, 48, 127, 1976.
265. **Pungor, E. and Umezawa, Y.,** *Anal. Chem.*, 55, 1432, 1983.
266. **Havas, J., Kecskés, L., and Erdélyi, J.,** *Hung. Sci. Instrum.*, 51, 27, 1981.
267. **St. Deutsch, Meier, P. Ch., Perriset, Ph., Pretsch, E., Simon, W., Clerc., J. Th., Gratzl, M., and Pungor, E.,** Microprocessor-Controlled Potentiometric Concentration Measurement of Ions in Blood Serum, Paper presented at the Conference on Computer-Based Analytical Chemistry, Portoroz, 1979.
268. **Růžička, J., Hansen, E. H., and Zagatto, E. A.,** *Anal. Chim. Acta,* 88, 1, 1977.
269. **Betteridge, D. and Růžička, J.,** *Talanta,* 23, 409, 1976.
270. **Pungor, E., Fehér, Zs., Nagy, G., Tóth, K., Horvai, Gy., and Gratzl, M.,** *Anal. Chim. Acta,* 109, 1, 1979.
271. **Thomas, J. D. R.,** in *Ion-Selective Electrodes, Vol. 4,* Pungor, E., Ed., Akadémiai Kiadó, Budapest, 1985, 213.

INDEX

A

- Absolute value, effect of, 27
- Absorbed ions, amount of, 94
- AC current, 4, 31
- Activation energy, 16, 39, 72, 83, 95
- Activity gradients, 53
- Activity independent slope values, 52, 54
- Activity level, 28, 39, 49—54, 82
- Activity step, method of, 2, 21—32, 34—35, 38, 44, 47, 77
  - decreasing vs. increasing interfering ion, 95
  - experimental techniques, 22—26
  - factors affecting shape of, 22
  - interfering ion, 91
  - measuring technique, 21—22
  - other sources of potential, 22
  - parameters affecting transient function of ion-selective electrodes, 26—31
  - perturbing signal, 21
  - principle of, 21—22
  - small, 77
  - two-ion range, 82
- Additional diffusion layer, 43
- Additives, 74, 76
- Adsorption/desorption processes, 20, 63, 77, 90—105
  - boundary conditions for, 99
  - differences in surface activity due to, 90—104
    - qualitative interpretation, 90—97
    - quantitative interpretation, 97—104
  - experimental proof for, 103
  - potential changes defined by, 105
  - validity, 105
- Adsorption isotherm, 97, 98, 101, 103
- Adsorption kinetics, 3, 8
- AgBr, 9
- AgCl, 9, 19, 31, 88, 97
- AgI-based iodide electrode, bromide interfering ions, 40, 90
- AgI pellets, 103
- Ag_2S, 9, 19, 32, 78, 117—119
- Ag_2S/AgX-based electrodes, 116
- Ag_2S-based silver ion-selective electrodes, 19, 114
- Alkali ion-selective glass electrodes, 84
- Almost zero-slope, 64
- Amplitude, 4, 31
- Analytical problems, application of electrodes for, 80
- Angular frequency, 4
- Apparent concentration, 58
- Apparent coverage factors, 86—87, 105
- Apparent ionic activities, 56
- Apparent selectivity coefficient, 86—87, 89, 104
- Aqueous boundary layer, 43, 69
- Assumptions used for mathematical descriptions, 98
- Asymmetric potential peaks, 123
- Average diffusion coefficient, 90

B

- Bathing solution activity, 8
- Biological samples, 121
- Boundary conditions, adsorption/desorption model, 48, 54, 69, 99, 100
- Bulk activity, 69
- Bulk electrode resistance, 9
- Bulk sample activities, 86
- Bulk selectivities, 83

C

- Cadmium (II) ions, 103
- Calculated response time curves, 45
- Capacitance, 4—6, 14—15, 60—61
- Carrier membrane electrodes, 69—71
- Cation-carrier complex, 68
- Cation-selective glass electrodes, 84, 114—115
- Cell assembly, 34
- Cell geometry, 47
- Cell voltage, 33, 39, 85
- Charge carriers, 8
- Charge distribution model, 17
- Charge transfer, 11—13, 17, 34, 51, 60—62, 77
- Charging current density, 61
- Chemical composition, 30
- Chemical homogeneity, 8
- Chemisorbed interfering ions, 97, 98
- Chloride electrode, 20, 66
- Circuits, 5, 7
- Complex impedance, 4, 6, 7
- Concentration, 9, 16—17, 27, 45—46, 53, 57, 79
- Conditioning, 28
- Conductivity, 4
- Consecutive overall reaction, 63
- Consecutive partial processes, 78
- Consecutive reaction model, 49, 62—64
- Constant membrane composition, 35, 39, 41, 54, 70, 81, 113
- Container, 23
- Controlled hydrodynamic conditions, 22
- Copper (II), 52
- Coverage factor, 87—88
- Critical activity, 50, 51
- Crystallization, 51, 66, 77
- Current densities, 11, 55—56
- Current-voltage curves, 19
- Curve fitting, 38, 65
- CuS-based copper(II) electrode, 103

D

- Dehydration, 62, 63
- Derivation, boundary conditions for, 100
- Desolvation reaction, 62

Desorbed ions, amount of, 93, 94, 101
Desorption parallel to adsorption, 97
Detection limit of sensor, 49
Development, 1—2
Dielectric relaxation, 9
Differential quotient, 52, 109
Different selectivities, layers of, 90
Diffusion coefficient, 43, 48, 69, 99
Diffusion layer thickness, 39, 43, 54, 69
Diffusion model, 35, 44, 48, 51, 68
Diffusion potential, 15, 16, 22, 28, 41
Diffusion processes, stagnant solution film, 25
Diffusion rates, 100, 118—119
Diffusion through a stagnant layer, 35—60, 67, 88, 109
  adhering solution film, 113
  hydrodynamic solution layer, 115
  ion-sensing membrane, 68—77
  limits of model, 44—54
    effect of activity level, 49—51
    effect of activity ratio, 51—54
  membrane bulk, 81, 115
  multielectrode model, 54—60
  resistance toward, 72
Digital simulation, 32
Dipping technique, 22—23
Dirac-delta function, 99
Direction of activity, effect of, 28
Dissolution, 3, 51, 66, 77, 79, 87
Distortion, 7, 32, 115
Distribution, 9
Donnan equilibrium, 1
Double diffuse layer capacitance, 9, 14—15, 60—61
Dynamic behavior, neutral carrier membranes, 69
Dynamic properties, 3, 19—21, 31, 35, 121—124
Dynamic response, 3, 17, 39, 72,107—108, 113
Dynamic response curves, 19—21, 29, 39—40, 48, 64

E

Electrical properties, 3, 60
Electrochemical cells, 5, 21—22, 107, 108
Electrochemical kinetic studies, 11
Electrochemical reactions, 3—4
Electrochemical sensor, selectivity of, 17
Electrode kinetic model, 16, 20, 105
Electrode potential, 11, 113
Electrode surface, 23, 25, 59, 95—96
Electrolysis, interface of two immiscible electrolyte solutions, 20
Electronic unit, transient response of, 107
Elementary electrode, 54
EMF response vs. time profile influence of membrane properties on, 73, 75
Empirical constants, 66
Empirical equations, 77
Energy barrier concept, 60—62, 115, 118
Enzyme electrodes, 78
Equilibrium cell voltage, 34
Equilibrium constant, 69
Equilibrium electrode potential, 11, 66, 78, 108, 112—114
Equivalent circuit models, 5, 8, 34
E vs. time curve, fluoride electrode, 67
Exchange current densities, 9—17, 20, 31, 49, 60, 83
Experimental conditions, 45, 79, 107, 119
Experimental response time curves, 110
Experiment setups, 24—25, 55
Experimental time constants, different K^+-selective electrodes, 74
Experimental transient curves two-ion range, 99
Exponential functions, 39, 61, 62, 70, 79
Extraction kinetics, 3, 69, 72

F

Fast flow technique, 116
Fick's law, 35
First-order chemical kinetics, 62—64, 118
Flow analytical techniques, 2, 81, 121
Flow conditions, 80
Flow injection analysis, 121—123
Flow profile variations, 47
Flowrate, 22, 26, 43, 50, 54, 60, 67
  dependence of transient function, 29
  effect of, 29—30, 44, 90, 94
  profiles, 45—46
Flow-through techniques, 22, 113, 116
Fluoride electrodes, 65—67, 116
Frequency, 8

G

Galvanic cell, 33—34, 58
Galvanostatic current step method, 13—15, 61
Galvanostatic perturbation, 14
Galvanostatic polarization current density, 61
Gas electrodes, 43, 78
Gas-permeable membrane, 43
Generally acceptable method of determination of response, 78
General purpose glass, 7
Generation rate, 8
Geometric bulk, 9
Geometric surface, 59
Glass electrode, 8, 9, 16, 23, 35, 39, 54, 60, 83, 86, 90, 96, 105

H

High precision monitoring, 121
High resistance region, 9, 43, 75, 115
H^+ ion exchange, 8
History and development, 1—2
Homogeneous phase with two boundary layers, 16

Hydrodynamic boundary layer, 35, 45, 47, 54
diffusion across, 63, 88, 118—119
ion diffusion within, 59, 115
rate-determining process, 44
thickness of, 39, 47—48
turbulent flow profile, 60
Hydrodynamic conditions, 27—30
Hydrogen ion-selective glass electrodes, 9, 21, 84
Hyperbolic functions, 64, 66—67, 70, 77, 116, 118
Hypothetical calibration graph, 82
Hypothetic current voltage curves, 18

I

Immersion technique, 22, 23
Impedance characteristics, 4, 8
Impedance method, 3—10, 31—32
impedance plane plots, 4—11, 32
application for ion-selective electrode studies, 8
capacitance, 5, 6
circuits, 5
defintions, 4—8
distortion, 7
factors determining, 8
information obtained from, 8—11
resistance, 5
Impedance spectrum, 6—7
Impedance studies, 31
Indicator electrode, 107, 113
Inductance, 4
Inhomogeneities, 84
Initial slope values, 49, 52, 54, 74, 109
Injection method, 22—23, 25
Interface of two immiscible electrolyte solutions (ITIES), 16, 20
Interfacial reactions, kinetics of, 60—68
energy barrier concept, 60—62
first-order chemical kinetics, 62—64
second-order chemical kinetics, 64—68
Interferences, 11, 27—28, 88, 91, 94
Interfering ions, 27—28
absence of, 33—80
diffusion through stagnant layer, 35—60
diffusion within ion-sensing membrane, 68—77
kinetics of interfacial reactions, 60—68
unified models for transient functions, 77—80
AgI-based iodide electrode, 90
membrane phase, 83
transient potentials, 81—105
comparison of different models, 104—105
differences in surface activity, 90—104
qualitative interpretation, 90—97
quantitative interpretation, 97—104
segmented membrane model, 84—90
Iodide-selective electrode, 49
Ion activities, 121
Ion-exchange membrane electrodes, 43, 89, 96
Ion-exchange process, 20, 63, 75, 76, 87
Ion-selective electrodes, 1—2, 16—20, 107—108, see also specific topics
Ion-selective membranes, 4, 34—36
composition, 27
inhomogeneities, 84
mixed potential of, 55
structure of surface, 95
Ion-sensing membrane, diffusion within, 68—77
Ionic mobilities, 83
Ionic strength, 28, 30
Ionophore-based electrodes, 80—81, 113, 115
Isotope distribution, 3
Isotope-exchange reaction, 19

K

K^+-selective carrier membrane electrode, 72—75
Kinetic model, 17
Kinetics of dissolution, 20
Kinetic studies with ion-selective electrodes, 11

L

LaF_3, 9
LaF_3-based fluoride ion-selective electrode, 78, 86
Laminar flow, 45, 46, 48
Lanthanum fluoride-based fluoride ion-selective electrode, 65
Leaching, 28
Lead (II) ions, 103
Ligand (carrier) concentration, 76
Linear flow rate, 26, 47
Linear perturbation, 8
Linear range, 38, 61—62, 77
Lipophilicity, 13, 20, 74—75, 77
Liquid ion-exchanger-based nitrate and calcium ion-selective electrodes, 54, 83, 90, 104, 114
Liquid ion-exchanger membranes, 9, 32, 35, 86, 105
Long-time behavior, 36
Lower detection limit, 43, 107

M

Mathematical formulation, 34, 115
Measured response time curves, 45
Measuring amplifier, 22
Measuring electronics, 113
Measuring setup, 26, 29, 32, 48, 54—55, 107
Membrane boundary, 69
Membrane bulk, 8, 9, 30—31
Membrane electrode, 34
Membrane matrix, 76
Membrane phase composition, 68
Membrane properties, 80
Membrane resistance, 13
Membrane-solution interface phenomena at, 3
Membrane transport experiments, 3
Metallic electrodes, 55, 58
Metallic silver, 119

Microelectrodes, 22, 34
Millisecond range, 114—115
Mixed kinetics, 119
Mixed potential, 31, 54, 56—58
Mixed potential ion-selective electrode theory, 16—20
Mixed potential of ion-selective membranes, 55
Mixing chamber, 26
Mobility, 9
Models, 69, 83, 90, 104—105, see also specific types
Multielectrode model, 54—60
Multilayer membrane model, 89
Multiparametric curve fittings, 79

N

NaTPB, 74—76
Nernst diffusion layer, 45, 47
Nernst equation, 33, 63—64, 67
Nernstian boundary layer, 45, 47—48
Nernstian response, 33, 49
Nernstian slope, 33
Nernst-Planck equation, 16
Neutral carrier-based ion-selective liquid membrane electrodes, lifetime of, 21
Neutral carrier ion-selective electrode, 8, 70, 121
Neutral carrier membrane electrodes, 68, 69, 71, 75
Nicolsky equation, 98
Nonelectrochemical methods, 19
Nonglass ion-selective electrode, 62
Noninterfering cations, 42
Nonmonotonic dynamic response curves, 90
Nonmonotonic potential overshoot-type transient functions, 84, 97
Nonmonotonic potential responses, 86, 103
Nonmonotonic transient electrode response, 98
Nonmonotonic transient functions, 11, 42
  qualitative interpretation, 90—97
  quantitative interpretation, 97—104
Nonmonotonic transient signals, 17, 83, 90, 93, 96, 98, 105
No potential response, 86
Normalized initial slope, 49—50, 74, 108

O

Ohmic potential drop, 13, 16
Ohmic resistance, 5, 32
Overall electrode process, 28
Overall electrode reaction, 78, 81, 116
Overpotential, 12, 17
Overshoot, 82, 92, 95, 99, 103

P

Parallel processes, 77—78
Partial processes, 78—79, 81
Partition parameter, 69
Perturbation, 3, 8, 14, 31
Perturbing signal, 21
pH glass electrodes, 25, 114—115
Phase angle, 4
Phase boundaries, 8, 16, 62, 118
Physical uniformity, 8
Physicochemical constants, 51
Physicochemical models, 63
Point-like electrodes, 59
Poisons, 28
Polarization resistance, 12
Polarization studies, 3, 11—20, 31
Polishing, 28
Potassium ion-selective electrodes response time curves, 76
Potential vs. time relationship, 13, 47, 64
Potential vs. time transient functions, 77
Potential determining ions, 8, 17, 35
Potential excursions, 83
Potential oscillations, 80
Potential response of Ag_2S electrode, 78
Potential response of cell, 16, 34
Potential response vs. time profiles, 32, 43, 72, 77
Potentiometric cell, input and output signals, 123—124
Potentiometric conditions, 39
Potentiometric selectivity coefficient, 81, 86
Potentiometric sensors, 32, 107, 113, 122
Practical applications, 108—109
Prandtl number, 45—46
Precipitate-based electrodes, 20, 23, 43, 54, 70, 83, 88, 90, 104, 105, 114—115
Precipitate-based membranes, 35
Precipitate ion-exchange reaction, 89
Preequilibrium current, 83, 105
Preparation technology, 30
Pretreatment of electrode surface, effect of, 91
Primary and interfering ions, 17
Primary ion activity, 58, 65, 67, 80
Primary ion concentration, 27—28, 51, 83, 88, 91, 101, 105
Process identification, 10
Pseudocapacitance, 10
Pumping frequency, 30
PVC, 20, 74, 76

R

Radioactive tracer techniques, 19
Radiochemical methods, 20
Range of some minutes, 115
Rapid potential response, 86
Rate constant, 62, 64
Rate controlling, 11, 51, 116
Rate-determining process, see Rate-determining step
Rate-determining step, 34, 44, 54, 60, 62—63, 67, 77, 81, 88, 89, 94, 105
Rate of growth, 66
Rate of reaction, 3

Reaction mechanisms, 11
Reaction orders, 11
Real and imaginary components, 4
Relaxation techniques, 3, see also Transient techniques
Reproducible flow conditions, 29
Required analytical accuracy, 110
Residence time, 123
Residual relative deviation in cell voltage, 39
Residual sum of X Squares (RSS) definition, 40
Resistance, 5, 9, 57, 72
Resistive surface layers, 9
Response functions, 3, 54, 88, 121
Response time, 29, 32, 39, 42, 48, 51, 64, 68, see also Transient function
  carrier membranes, 69
  determination and definition of, 107—119
    advantages of, 110
    definition, 108—112
    dynamic response of ion-selective electrode, electrochemical cell, 107—108
    estimation of the equilibrium potential, 112—114
  differential quotient-based definition, 112
  indicator electrode, 107
  influences upon, 72
  measuring conditions, 23
  new slope-based definition, 111
  selected examples, 114—119
    interpretation of data, 116—119
  variation of, 28
Response time curves, 49, 50, 52, 59, 74, 76, 78, 80, 111, 115, 117—118
  corrected initial slope, 48
  fluoride electrode, 65
  silicone rubber iodide-selective electrode, 42
  slope of, 52
  two-ion range, 82
Reynolds number, 46—47
RMSD, 40
Roughness of membrane surface, 59
RSS, see Residual sum of X Squares

S

Sand equation, 15
Schmidt number, 46
Second-order chemical kinetics, 64—67, 115
Segmented membrane model, 16, 84—90, 104—105
Selectivity coefficient, 87, 107
Semicircles, 7
Sensitivity, 11, 121—122
Sensors, 121
Sequence of partial reactions, 11
Short response and long life time, sensor of, 21
Short-time behavior, 36, 70
Silicone rubber (SR), 76
Silicone rubber-based electrodes, 42—45, 74
Silicone rubber membranes, 74
Silver bromide-covered silver iodide-based electrodes, 20, 97
Silver chloride membrane electrode time-dependent response, 89
Silver halide-based electrodes of the second kind, 114
Silver iodide-based electrodes, 48, 58, 60, 95, 100
Silver metal electrode, 55, 60, 116, 117
Silver-selective electrode, 59
Silver sulfide-based silver ion-selective electrode, 65
Simplified equivalent circuit for membrane electrodes, 13
Slope values, 50, 54, 108, 111
Slow potential response, 85—86
Slow surface modifications, 95
Sluggish electrodes, 39, 43, 78, 84
Sodium ion-selective glass electrodes, 66, 115
Sodium tetraphenylborate, see NaTPB
Softener, 74
Solid-state membrane approach, 35, 105
Solubility products, 87
Solution-membrane interface, 32
Solvent properties, 54
Space charges, 8, 9, 16
Square root of time dependence, 70
SR, see Silicone rubber
Stagnant solution layer, 35
Standardized conditions, 107
Steady-state flow conditions, 47, 113, 121
Stirrer geometry, 23
Streaming potential, 22, 28—30
Sub-Nernstian slope, 11
Successive potential overshoots, 103
Sunken semicircles, 7
Surface activities, 35, 63, 66—67, 87—88, 90, 98
    change in time, 37—38
  differences due to adsorption/desorption processes, 90—104
    qualitative interpretation, 90—94
    quantitative interpretation, 97—104
  variation, 101
Surface area, 54, 60, 88
Surface change model, 95
Surface conditions, 30
Surface coverage, 87—88
Surface etching, 28
Surface morphology, 39
Surface resistance, 9—10, 60
Surface roughness, 48
Surface selectivities, 83
Surface wetting, 39
Surfactants, 28
Switched wall-jet cell, 1, 25
Swollen boundary phase, 8

T

Temperature, 8, 96—97

Theoretical EMF response vs. time profiles, 71
Theoretical potential response vs. time profiles, 64, 109, 112
Theoretical selectivity factor, 86
Thickness, 30, 35, 39, 46—48, 54, 60, 90
Three-segmented potential model, 16
Time constant, 6, 8, 9, 14, 31, 36—37, 39, 43, 70—73, 79, 83, 95, 107
Time dependence, 34, 37, 54, 70, 86, 102, 104—105
Time-dependent response, 89, 100
Time of mixing, 23
Total current density, 11
Transfer rate, 8
Transformation of ion-selective membrane, 88—89
Transient function, 22, 29, 32, 34, 54, 60, 62—68, 81, 84, 105, see also Response time
  direction of concentration change, effect of, 58, 63
  electrochemical cell, 108
  ion-selective electrodes, parameters affecting, 26—31
    hydrodynamic conditions, 28—30
    interferences, 28
    membrane bulk and surface parameters, 30—31
    primary ion concentration, 27—28
  long-time behavior, 36
  potentiometric system, 107
  precipitate-based electrodes in dilute solutions, 70
  preselected point of, 108
  PVC membranes within NATPB, 74
  short-time behavior, 36
  SR membranes, 74
  two-ion range, 89
  unified models for, 77—80
Transient potentials, 42, 90, see specific topics
Transient response, 43, 66, 107
Transient signals, 49—50, 60—61, 82, 83, 92, 94—97, 102—104
Transient techniques, 3—32
  activity step method, 2, 21—31
    experimental techniques, 22—26
    parameters affecting, 26—31
    principles, 21—22
  comparison, 31—32
  definition of, 3
  dynamic properties, 3, 20—21
  impedance methods, 3—10
    application of, 8
    impedance plane plots, 4—7
    information obtained from 8—10
  information content of, 3
  membrane transport experiments, 3
  polarization studies, 3, 11—20
    methods, 11—16
    mixed potential ion-selective electrode theory, 16—20
  selection of, 3
Transition time, 15
Transmission line, 7
Transport, 8, 35, 77
Tube diameter, 30
Turbulent flow, 45—48, 60
Two-ion range, 81, 82, 89, 92, 95, 97, 99, 123
Two-layer membrane model, 84

U

Unified models for transient functions, 77—80

V

Valinomycin-based K^+-selective electrode, 9, 75, 114
Velocity profile, 46
Viscosity, 22, 28, 39
Viscous boundary layer, 46—48

W

Wall-jet cell arrangement, 24, 26, 47, 60
Warburg behavior, 36
Warburg diffusion, 7, 9, 10
Warburg resistance, 9
Working mechanism of all types of ion-selective electrodes, 11

Z

Zero initial slope, 109